TABLE OF CONTENTS

✗. very imp

ACKNOWLEDGMENTS

Waste Management, Inc. and the law firm of Piper & Marbury have produced this book only through the efforts of numerous people. The effort was spearheaded by William Y. Brown, Director of Environmental Affairs of Waste Management, Inc., and Mary F. Edgar, a partner in the Environmental Department of Piper & Marbury's Washington, D.C. office. Significant contributions to this book's success were made by Stephen R. Linne and Stephen N. Herbst of Chemical Waste Management, Inc. and Cathy Barney of Waste Management of North America, Inc., as well as Andrew S. Miller and Norman L. Rave, Jr., associates at Piper & Marbury. Finally, special thanks are extended to Beth Hopkins and Sandy Donahue of Piper & Marbury for their editorial, paralegal, and secretarial assistance.

CHAPTER ONE

Overview

This book is about the policy and practice to reduce the amount and toxicity of solid waste produced in the United States. We call this *waste reduction*, which we define to be the sum of "source reduction" and recycling. Our society generates almost incomprehensible volumes of waste. Improper and inefficient management of that waste produces both environmental and economic costs. From the efforts of industry, regulators, and concerned members of the public who grapple with these problems has emerged a general consensus that gives waste reduction a preferential status in a hierarchy of waste management practices.

THE SOLID WASTE STREAM

By "solid waste," we mean the term as defined in the Solid Waste Disposal Act, typically referred to as RCRA.[1] Under RCRA, solid waste is, with certain specific exceptions, "any garbage, refuse, sludge,...and other discarded material," including any "solid, liquid, semisolid, or contained gaseous material."[2] That definition encompasses a wide range of things: from sewage sludge to candy wrappers to liquid chlorinated organic solvents. Despite the breadth of this definition, many of the polluting by-products of modern society are not solid wastes regulated under RCRA.

Many air pollutants, for example, are regulated under the Clean Air Act but not RCRA. Emissions from coal-fired power plants annually include millions of tons of sulphur dioxide, a leading pollutant of concern in the ongoing debate on acid precipitation. Enormous quantities of nitrogen oxides, carbon monoxide, and assorted hydrocarbons are emitted from the tailpipes of cars, trucks, and buses. More than 1 million tons (2.4 billion pounds) of 108 specific toxic chemicals and twenty classes of toxic chemicals were emitted into the air by large manufacturers in 1987, according to reports filed under amendments to the Superfund legislation.[3]

Much pollution discharged into water also is not regulated as solid waste. For example, each day sewage treatment plants pipe more than 9.2 billion gallons of treated wastewater containing various pollutants into our nation's water bodies.[4] This volume does not include discharges with contaminant levels exceeding Clean Water Act standards. These discharges are in fact wastes from our homes and our factories, regulated under the Clean Water Act, but exempted from RCRA's requirements. Separate federal regulations also apply to about 180 million tons of dredge spoils dumped annually in estuaries or at sea and about 7 million tons of wet sludge dumped at sea each year.[5]

Despite these congressionally defined diversions from the stream of RCRA solid wastes, the remaining flow is massive. The United States Environmental Protection Agency (EPA) estimates total annual nonhazardous solid waste generation at 11.4 billion tons,[6] not including agricultural solid wastes of up to 1.4 billion tons per year.[7] Hazardous waste accounts for about 0.27 billion tons.[8]

Brought down to earth, these statistics mean that about fifty tons of solid waste are generated each year for each person living in this country.[9] On average, each person is responsible for, or at least associated with, generation of a little more than 300 pounds of solid waste each day—about twice the average body weight. About 190 of these pounds are assorted industrial wastes, seventy-one pounds are oil and gas wastes, and thirty-five pounds or more are mining wastes.[10] Hazardous waste makes up six to twelve pounds. Trash, or municipal solid waste (MSW), amounts to only three to five pounds of the total. Although it is a mere rivulet in the total solid waste stream, MSW is getting much attention, presumably because we see it every day and generate it directly (e.g., throwing a candy wrapper into a trash can) instead of being in a position several steps removed (e.g., using a personal computer, the manufacture of which generated a sludge containing toxic metals).

THE WASTE MANAGEMENT HIERARCHY

Environmental and economic concerns lead logically to the conclusion that waste reduction must be promoted.[11] These concerns have prompted the EPA, members of Congress, environmental organizations, profit-making companies, and others to endorse the concept of a hierarchy of preferences among waste management techniques.[12] This hierarchy, which is based upon the apparent effectiveness in preventing pollution, begins with source reduction and recycling. The following is consistent with many views of the hierarchy:

Source reduction
Recycling
Treatment
Land disposal
Managed dispersion[13]
Unmanaged dispersion[14]

The hierarchy begins with waste reduction. If waste is not made at all, less pollution will enter the environment. Similarly, most recycling practices will produce less pollution of concern than, for example, dumping the same wastes into the ocean. The hierarchy has limitations, however. Some wastes rich in heavy metals, for example, may produce undesirable air emissions if burned for energy recovery. Therefore they may be more prudently managed through stabilization or other forms of treatment, followed by landfill burial of the treated residue.

The hierarchy of practices for dealing with waste often is truncated, ending with land disposal. The bulk of pollutants generated in this country is managed by dispersion into the air or water, however, and the bulk of nonhazardous solid wastes is put into unlined waste ponds that cannot be said to contain wastes to a high degree. Furthermore, some of the nastier municipal discards such as lead-acid batteries and tires often evade both recycling and disposal facilities, thus magnifying their potential adverse effects on the environment. For example, as little as 3 percent of nonrecycled lead-acid batteries may be sent to landfills. The rest just go over the fence, into junkyards, or somewhere else.[15] It is unfortunately true that the waste management hierarchy actually in use in this country includes practices that are unsafe. Regardless of what sits at the bottom of the hierarchy, waste reduction is appropriately positioned on top.

Environmental concerns are not the sole reasons for the advance of waste reduction. Consider MSW. There are now about 6,000 active landfills in the United States, but 45 percent of them are expected to close within five years.[16] Siting of new landfills and incinerators is typically difficult and fraught with delay, and the cost of MSW disposal has risen sharply. Thus, economics demand an increase in recycling and source reduction of MSW. The same economic forces are also at work with respect to other solid wastes.

Recycling probably has the greatest potential for reducing municipal solid waste over the next decade. The process of diverting material from the waste stream for reuse is not the limiting factor in promoting recycling. Instead, the success of recycling programs depends on finding buyers for diverted goods, or at least finding institutions that will take the goods for reuse without charge or at a cost that makes recycling economically attractive. Unfortunately, the markets for some materials—newsprint and compost in particular—are weak.

Source reduction has great potential for reducing all kinds of solid waste over the long term. It can be accomplished through reduction in either waste volume or toxicity. Source reduction involves far more complexities than recycling, however, and policy approaches to it reflect that difference.

THIS BOOK

This book begins with a characterization of solid waste streams, emphasizing hazardous waste and municipal solid waste. The next two chapters address how the reduction of industrial wastes and MSW can be accomplished. The section on the means for industrial waste reduction draws upon the substantial experience of chemical waste generators in reducing waste through source reduction and recycling and also addresses waste reduction services offered commercially. The MSW section discusses thinking, which is still in an embryonic stage, on source reduction, and then addresses the opportunities and means for recycling items extracted from trash. This has become a priority of cities, states, industry, and environmentalists across the country.

The book closes with a discussion of waste reduction policy. Existing federal and state requirements for reducing industrial wastes are addressed first, followed by a discussion of recent federal legislative initiatives in the area. A similar review follows for MSW reduction policy.

Waste reduction is a hot topic; it has become a Holy Grail. Waste reduction warrants more than theological debate, however. This book puts some flesh on the subject's bones. If the book's readers develop a feel for the subject, an appreciation of the enormous potential savings for society and individuals presented by waste reduction, and some direction about the variety of ways in which waste reduction can be accomplished, the authors have accomplished their objective.

CHAPTER TWO

Components of Solid Waste

The roughly 12 billion tons of solid waste generated annually in the United States are managed in 227,000 facilities located throughout the country.[17] These wastes and their management are described below under two broad categories: (1) municipal solid waste (i.e., trash) and (2) other wastes.

MUNICIPAL SOLID WASTE

Americans generate approximately 180 million tons of municipal solid waste annually. Today, approximately 73 percent is buried in landfills; 14 percent is incinerated; and 13 percent is recycled.[18] The major components of this waste stream, by weight and amount recycled, are summarized in **Table 1** (at p. 18 *infra*) and described below.

Paper

Paper, which constitutes approximately 34 percent by volume of the municipal solid waste (MSW) stream, is the largest materials category. It includes newspapers, books and magazines, office papers, corrugated cardboard, and other commercial paper.

Nationwide, about 4 million tons of newspapers are recycled every year, almost all collected from various kinds of source separation programs. Old newspapers are recycled principally into paperboard or new newsprint. Recycled newspaper can also be used to create cereal boxes, writing pad bases, wall board, corrugated containers, tissue paper, and bedding for animals.

Glass

Glass constitutes approximately 2 percent by volume of the MSW stream. The tonnage of glass in the waste stream comprises mostly containers, including beer, soft drink, wine, and other liquor bottles.

For recycling purposes, glass containers are classified into three categories according to color: clear, green, and amber. The clear bottles account for about 70 percent of all glass containers. In the glass manufacturing business, up to 15 percent of waste can replace all or part of the mineral raw materials. Glass containers are collected and remelted with raw materials to create new glass bottles and jars or fiberglass.

Metals

Metals constitute approximately 10 percent of the MSW stream. Metals in the municipal solid waste stream include beverage cans, food and other cans, foil, and closures.

Beverage cans are the largest source of recycled aluminum. Small amounts of aluminum occur as foil and trays. Although the fraction of aluminum in the waste stream is low, it is a valuable material because of its high price as scrap. In most cases, recycled aluminum is used to produce new cans.

Ferrous scrap recovered from MSW includes steel cans and white goods. Recycled steel cans are used to make new steel from manufacturing, and the tin removed from the outer coating of cans is used to make new tin products.

Plastics

Plastics constitute 9 percent of the MSW stream by weight and approximately 20 percent by volume. Plastics have increased steadily in the waste stream since 1960, when they were less than 1 percent of the waste stream, and in 1984, when plastics accounted for approximately 6 percent of gross discards.

Three kinds of plastics that are commonly found in residential and commercial waste can be readily recycled. These are polyethylene terephthalate (PET), which is used to make plastic soft drink bottles; high density polyethylene (HDPE), which is used to make plastic milk and juice containers; and low density polyethylene (LDPE), which is used to make film products such as shopping bags and garbage bags. Recycled plastics are used in flower pots, drainage pipes, toys, traffic barrier cones, carpet backing, and fiberfill for pillows, ski jackets, and sleeping bags.

Rubber and Leather

Rubber and leather constitute approximately 3 percent by weight of the MSW stream. This category includes rubber tires, hoses and belts, and fabricated rubber products. Approximately 2 million tons of tires are thrown away each year and are very expensive to dispose of because they are bulky and take up valuable landfill space. They can be used in asphalt pavement, playground rubber mats, railroad crossings, and industrial fuel.

Textiles and Wood

Textiles constitute approximately 3 percent of the MSW stream, and wood about 4 percent. Trees, branches, and stumps are often recycled and sold as wood chips or firewood.

Food Waste, Yard Waste, and Miscellaneous Inorganic Waste

Food wastes constitute approximately 7 percent of the MSW stream. Yard wastes, which include grass clippings and brush, constitute approximately 18 percent by weight of the municipal solid waste stream. Yard waste recycled into compost can be used to enrich soil. The remaining miscellaneous inorganic wastes, which represent 2 percent of the MSW stream, include mostly stones and dirt.

OTHER SOLID WASTES

The bulk of the solid wastes in the United States is generated by industry: almost 8 billion tons of industrial nonhazardous waste and approximately 275 million tons of hazardous waste. Other significant waste streams are construction and demolition waste, municipal sludge, ash from municipal waste incinerators, and agricultural waste.

Industrial Nonhazardous Waste

A total of 7.6 tons of various industrial wastes is discharged into 15,000 waste ponds, 5,000 waste piles, 4,000 land application units (LAUs), and 3,000 landfills.[19] The wastes come from at least twenty-two different industries, led by industrial organic chemical manufacturers.[20] About 96 percent of the waste is disposed in waste ponds.[21] Only about 4.7 percent of these waste ponds have a synthetic liner; only 17.4 percent have some kind of natural liner; and only 5.5 percent have a leak detection system.[22] More than half of these waste streams contain relatively high levels of heavy metals and organic constituents, some of which are extremely toxic.

Not counted in this 7.6 billion tons are the wastes from three industries producing high volumes of nonhazardous wastes. Two to four billion tons of brines and drilling muds from oil and gas exploration and production are discharged into 126,000 waste ponds and 700 LAUs.[23] Synthetic liners are used in only 2.4 percent of the ponds; only 27 percent have some kind of natural liner; and only 1.1 percent have a leak detection system.[24] These wastes are largely water, containing substantial amounts of chlorides, barium, sodium, and calcium, as well as chromium, benzene, and other toxic chemicals.[25]

The mining industry annually discharges more than 1.4 billion tons of wastes into 20,000 waste ponds.[26] These wastes are produced by activities such as crushing, screening, washing, and flotation. Only 1 percent of the mining waste ponds have a synthetic liner; only 4.4 percent have some kind of natural liner; and only 1.7 percent have a leak detection system.[27] These wastes contain heavy metals, sulfate, sodium, potassium, and cyanide.[28]

The third large category of industrial nonhazardous wastes is utility wastes. In 1984, coal-fired electric power plants produced 69 million tons of ash and 16 million tons of flue gas desulphurization wastes.[29] Approximately one-fifth of these wastes are being recycled.[30] The remainder goes to surface impoundments and landfills.[31]

Hazardous Waste

Between 250 million and 275 million tons of waste regulated as hazardous under RCRA are generated in the United States each year.[32] Process wastewaters represent by far the largest component of this volume. Ninety-six percent of the annual industrial hazardous waste generation is managed by the generator in captive, on-site processes.[33]

Generation has not grown significantly over the past three to five years, nor has it declined. Several offsetting factors have affected generation, however. Although waste reduction has become more attractive as costs of waste management have increased, new wastes and generators continue to be added to the regulated community. Economic growth has created a potential increase in waste generation, much of which has been mitigated by more efficient technology, and thus waste production is reduced at the source.

Hazardous wastes fall into four broad categories: organic wastewaters, organic nonwastewaters, inorganic wastewaters, and inorganic non-wastewaters. Organically contaminated wastes generally lend themselves to incineration or secondary fuels programs because of their high heat release value (BTUs). Some of these wastes may also be recovered for use as solvents. Inorganic wastes are generally contaminated with metals or are corrosive and, except for metals recovery through thermal separation, are generally not suitable for management by thermal technologies. The more aqueous a waste is, the less likely it is to have an economic heat value (from organics) or metal value (from inorganics). Each of these categories is described below. In addition, we discuss a class of hazardous waste that is defined by those who generate it—hazardous waste produced by small quantity generators.

Organic Wastewaters

Organic wastewaters represent 18 percent (approximately 45 million tons) of the hazardous waste stream. These are primarily aqueous wastes with at best a low heat value. Virtually all of this waste is managed on-site in deepwells or in wastewater treatment plants.[34] The cost of managing these wastes tends to be relatively low on a per unit basis, ranging from $.005 to $0.05 a gallon. Commercial management of these wastes is available but at a higher cost (roughly $0.05 to $0.50 a gallon). Therefore, such off-site management is generally reserved for the more complex organic wastestreams.

The chemical, petroleum, and automotive industries are the leading generators of organic wastewaters. The primary contaminants in these wastewaters come from chemical processes, spent solvents, paint applications, and drilling muds.

The EPA promulgated new regulations in 1990 that will bring approximately 800 million more tons of wastes into the hazardous waste category.[35] Most of this additional waste will be in the form of organic wastewaters, primarily from the petroleum refining industry and the chemical manufacturing industry. Some 200,000 generators are expected to be affected by the regulation.

Organic Nonwastewaters

Organic nonwastewaters represent only 1 percent of the hazardous wastes generated, but account for a disproportionate share of the dollars that industry spends to manage its waste each year. The relatively high costs reflect the fact that these wastes are typically managed in RCRA-regulated incinerators at costs ranging from $0.20 a pound to $1.50 a pound.

Again, the primary generators of these wastes are the chemical manufacturing and petroleum industries. Significant volumes of organic nonwastewaters are also produced by the wood preserving industry. The table below gives an overview of organic nonwastewaters by industry.

INDUSTRY	WASTE TYPES
Chemical	Spent solvent residuals, still bottoms, spent catalysts, treatment sludges, filter cakes
Petroleum refining	Slop oil emulsion solids, API separator sludge, DAF float
Wood preserving	Bottom sediment sludge from wastewater treatment
Pesticide/herbicide	Distillation residues, still bottoms, manufacturing treatment sludges

Polychlorinated biphenyls (PCBs) are a type of organic waste that merits a separate discussion because their disposal is regulated under the Toxic Substances Control Act rather than RCRA. PCBs are no longer manufactured, but still are found in many electrical transformers and capacitors. As these units are taken out of service and replaced, the PCB oils inside must be incinerated in TSCA-permitted incinerators or chemically broken down into nontoxic molecules. PCB-contaminated soil, debris, and electrical equipment carcasses may be landfilled, but only at TSCA-permitted facilities.

Inorganic Wastewaters

Inorganic wastewaters represent the largest category of hazardous waste, accounting for 75 percent of the annual generation. This 187 million tons is typically managed in on-site wastewater treatment units, deepwells, or surface impoundments. Like organic wastewaters, these management methods typically cost generators between half a cent to a nickel a gallon to handle. Again, off-site commercial management is available for the more complex of these wastestreams, but at a higher cost (roughly $0.10 to $1.00 a gallon or even higher).

These wastes fall into two main categories: metal-containing wastewaters and nonmetallic wastewaters. Metal-containing wastewaters are generated primarily by metal finishing industries, such as steel manufacturers, lead smelters, and mercury cell manufacturers. In addition, wash waters from electrical equipment manufacturing industries and fabricated metals industries also contribute significant volumes to metal-bearing wastewaters. Nonmetallic wastewaters mainly consist of corrosives and cyanides used in chemical manufacturing, pharmaceutical manufacturing, mining, and catalyst production.

Regulations restricting the use of deepwells and requiring stringent treatment of these wastes will tend to drive more of these waste streams toward more expensive off-site management technologies. Wastes bearing fairly small levels of cyanide will require stabilization prior to land disposal and wastes containing certain metals will require either metals recovery or stabilization and land disposal. Although the concentrations of contaminants in many wastewaters may not be high enough to require additional treatment, in a significant number of cases these additional steps will be necessary. The additional costs will likely triple or quadruple the cost of managing these streams.

Inorganic Nonwastewaters

Fifteen million tons of inorganic nonwastewaters are generated each year, accounting for 6 percent of U.S. hazardous waste generation. Currently managed primarily in landfills, these waste streams require significant (and expensive) pretreatment or recovery prior to land disposal. Typical costs for direct commercial land disposal today are $100 to $180 a ton. A treatment step such as stabilization at this time adds $50 to $80 a ton to the price. Most of this additional cost reflects the research and development effort required to arrive at "recipes" that adequately immobilize the contaminants of concern in a nonleachable matrix.

The other option for management of these wastes is metals recovery. Recovery tends to be an expensive process unless the recovered product has significant value on the open metals market. Therefore, this option is economically feasible for only a portion of metal contaminated wastes. An example is zinc recovery from electric-arc furnace dust, which returns zinc with a current market value of approximately $225 a ton.

The major generators of inorganic nonwastewaters are the steel industry, aluminum, zinc and lead smelters, electroplating processes, and metal fabricators. The largest streams in this category and the generating industries are given in the table below.

INDUSTRY	WASTE TYPES
Steel manufacturing	Electric-arc furnace dust, pickle liquor
Mining and primary	Metal contaminated sludges, mill tailings metals smelting
Electroplating	Electroplating sludges from wastewater treatment, tank bottoms
Metal fabricating	Electroplating sludges and tank bottoms, etching solutions, cyanide and metal contaminated sludges

Small Quantity Generator Hazardous Waste

More than 600,000 sources of hazardous waste (about 98 percent of generators) produce less than 1,000 kg of hazardous waste each month. These Small Quantity Generators (SQGs) contribute only 940,000 metric tons of waste annually, but are a major complicating factor in the regulation and management of hazardous wastes.

More than half of the SQG wastes are produced in motor vehicle maintenance. The other large contributors are metal manufacturing, printing and ceramics, and laundries. All SQG wastes have hazardous characteristics. Many kinds of wastes are involved, including used lead-acid batteries (nearly half of all SQG wastes), spent solvents, photographic wastes, and dry cleaning residues.

About 71 percent of SQG wastes, much of it lead-acid batteries, is recycled. Those wastes that are not recycled are discharged into sewers, treated in some fashion, or disposed of on land, but generally not in hazardous waste management facilities regulated under Subtitle C of RCRA. A total of 27,631 facilities that are not regulated under Subtitle C report receipt of SQG waste: 76 percent are waste ponds, 18 percent are landfills, and 6 percent are LAUs.[36]

Construction and Demolition Waste

About 31.5 million tons of construction and demolition wastes are produced annually. These wastes vary widely; they include mixed lumber, roofing and sheeting scraps, broken concrete, asphalt, brick, stone, plaster, wallboard, glass, piping, and other building materials. These wastes are placed into landfills or waste piles, which are often less stringently regulated than municipal solid waste landfills.[37]

Municipal Sewage Sludge

Some 7.6 million tons (dry weight) of sewage sludge are produced each year by publicly owned treatment works (POTWs). The sludge constituents vary, but often include cadium, copper, and zinc. Reportedly, 46.4 percent of the sludge is managed by "lagooning and landfilling." LAUs receive 25.4 percent; 20.3 percent is incinerated; and 6.16 percent is dumped at sea.[38]

Municipal Solid Waste Combustion Ash

In 1988, municipal solid waste incinerators produced an estimated 3 million to 8 million tons of ash. With the number of such incinerators expected to more than double by the year 2000,[39] the quantity of ash requiring management will increase dramatically. Today, 90 percent of this ash is landfilled, with a small portion being used for cover material, road construction, and in cement and concrete mixtures. There is growing pressure to find alternatives to landfilling because of the scarcity of landfill capacity, but also to employ special landfill design and operating practices because some of this ash has exhibited elevated levels of heavy metals, which are concentrated by the incineration process.[40]

Agricultural Wastes

Agricultural wastes are discharged into 17,000 waste ponds.[41] The EPA estimates that these ponds receive an "upper limit" of one billion gallons of agricultural waste a day (approximately 1.4 billion tons each year).[42] The discharges include animal wastes from feedlots and farms, crop production wastes, irrigation wastes, and collected field runoff, but do not include irrigation return flows or those wastes that are returned to the soil as fertilizers or soil conditioners.[43] Some 0.3 percent of these waste ponds have synthetic liners, 54.2 percent report some kind of natural liner, and 0.2 percent report a leak detection system.[44] These wastes commonly contain high concentrations of nitrates, pesticides, herbicides, and fertilizers. Agricultural wastes are exempted by RCRA from regulation as hazardous waste regardless of their constituents.

TABLE 1

MUNICIPAL SOLID WASTE IN 1986

Quantity Generated, Amount Recycled,
and Remaining Trash
(rounded to the nearest million tons and percent)[45]

Materials	Gross Discards	Amount Recycled	Remaining Trash
Paper & paperboard	72 (40%)	18 (25%)	54 (35%)
Yard wastes	32 (18%)	.5 (2%)	31 (20%)
Plastics	14 (8%)	.2 (1%)	14 (9%)
Glass	13 (7%)	2 (15%)	11 (7%)
Food wastes	13 (7%)	0 (0%)	13 (8%)
Ferrous metals	12 (7%)	1 (8%)	11 (7%)
Wood	7 (4%)	0 (0%)	7 (4%)
Rubber & leather	5 (3%)	.1 (2%)	5 (3%)
Textiles	4 (2%)	0 (0%)	4 (3%)
Aluminum	3 (2%)	1 (33%)	2 (1%)
Other	5 (3%)	1 (20%)	4 (3%)
Total	180 (100%)	24 (13%)	156 (100%)

CHAPTER THREE

Means of Waste Reduction

This chapter addresses opportunities for source reduction and recycling of, first, industrial waste and then, municipal solid waste. The discussion cannot cover all possible means of waste reduction but is intended to introduce the reader to the wide variety of waste reduction techniques available.

INDUSTRIAL WASTE REDUCTION EFFORTS

In the industrial sector, waste reduction is now being recognized as the most environmentally safe, and often the most cost-effective, waste management technique. For many years, industries have concentrated on meeting end-of-pipe permit limitations, giving little attention to upstream waste production. This attitude has been fostered by the development of relatively low cost, reliable treatment processes. As state and federal agencies move toward tighter restrictions, however, and begin eliminating some treatment practices (e.g., through the land ban regulations), industry must either look for new treatment technologies or closely examine the wastes they are producing and institute waste reduction measures.

The gains that can be made in this area are significant. The EPA has estimated that hazardous waste production in this country can be reduced by 15 to 30 percent, while the Congressional Office of Technology Assessment estimates the potential exists for a 50-percent reduction.[46] The 3M Corporation instituted its 3-Ps program (Pollution Prevention Pays) and realized a savings of $292,000,000 over an eleven-year period.[47] Other major industries have had similar successes.

This section of the book first describes the process by which a company can review its waste production to identify the possible means of reduction. We next discuss the most common means of achieving waste reduction, source reduction, and recycling. Finally, we provide three case studies in successful waste reduction.

Waste Reduction Reviews

Several key elements are necessary to the success of a waste reduction study. First and foremost is management support. Successful waste reduction studies examine all facets of an organization and ask how existing practices affect the generation of the waste. To obtain cooperation from the entire spectrum of departments and people within an industrial facility, a "waste reduction team" must have the support of the upper management. For example, in some cases, significant waste reduction has been realized by changing the way in which raw materials are purchased. Without upper management support, the potential for such a change may never even be identified if the Purchasing Department resists intrusion into its territory by the Waste Department.

After obtaining the required support, a waste reduction team must be formed. The team should include someone with a working knowledge of relevant environmental regulations, a person responsible for waste disposal, a representative from the firm's accounting and purchasing functions, a process engineer, a management representative and, possibly, an outside consultant.

An outside waste reduction consultant, who can look at the firm's current practices without bias, can be very helpful. The consultant can also develop recommendations that benefit the firm as a whole, rather than optimizing waste reduction in just a few departments. For example, a consultant recently completed a waste reduction study at a defense-related manufacturing facility that identified substantial waste disposal savings. In the process, the consultant discovered that the purchasing department was buying raw materials in 55-gallon drums, based purely on purchase price with no consideration of drum disposal costs. The empty drums were hauled away at a disposal cost of $20 for each drum. Meanwhile, another department was purchasing drums for waste disposal at a cost of $15 to $17 a drum. The consultant recommended (1) that the purchasing department account for the cost of drum disposal in all purchase decisions, thus creating an incentive to purchase material in returnable containers; (2) that the production department mark empty drums as to their acceptability for waste disposal; and (3) that the waste disposal department limit purchases of new drums by using suitable empty drums from the production department. The solution may seem very simple, but waste reduction most often is achieved through simple, commonsense approaches. Nevertheless, an outside person, the consultant, was needed to identify this potential for waste reduction across departmental lines.

Once management support is obtained and a waste reduction team is formed, the team should conduct a facility-wide waste reduction assessment. The goal of this assessment is to identify all major waste streams, their components, their origins, their disposal costs, the purchase and disposal quantities of each waste component, and the regulatory status of each component.

This process can be viewed as a mass balance analysis, first, around the entire plant and, then, around each process within the facility. If the reader is unfamiliar with the process of constructing a mass balance analysis, an introductory chemical engineering text should be consulted. Essentially, however, the assessment looks at what goes in and what comes out and identifies what happens in between.

Mass balance information can be obtained from a variety of sources short of actually measuring each waste flow and concentration. **Table 2** (at p. 39 *infra*) presents a list of sources of data for the mass balance analysis. Care should be taken to ensure that the mass flows are as accurate as possible. Assumptions as to flow rates, concentrations, or fugitive losses should be avoided. Also, the data used should reflect an extended period of time (typically a year) to ensure use of accurate average values.

A quantitative method to examine the amount of raw material ending up as waste is to calculate the actual material utilization ratio (AMU).[48] The formula is:

$$AMU\% = \frac{\text{minimum material needed}}{\text{actual material used}} \times 100.$$

If this ratio is less than 100, waste is being produced. The AMU ratio can also be very useful in tracking the performance of the waste reduction effort.

Other information will be needed to complete the waste reduction study such as cost, purchasing, and procedural information. **Table 3** (at p. 40 *infra*) identifies a variety of sources for this information. The time period for the necessary initial data-gathering effort will depend on the size of the facility and the accessibility of the data.

The next steps are the engineering and problem solving tasks. The specifics of these tasks are discussed in detail in the following sections of this chapter, which discuss waste reduction techniques; but briefly, waste reduction can be realized through source reduction and recycling. Each waste stream should be subjected to a detailed examination to identify ways to eliminate components of the waste stream through process changes, improved efficiencies, waste segregation, or waste concentration. If wastes or components cannot be eliminated, the next step is to determine whether and how the waste stream can be modified to make it more amenable to recycling. "Modified" does not necessarily mean treatment or processing.

The single most important element of a successful recycling effort is the elimination of contamination. Once a waste is earmarked for recycling, sources of contamination must be identified and removed or reduced as much as possible. Contaminants can be identified by working with a consultant and/or the potential receiver of the recyclable waste stream. A generator's relationship with the recycler of a waste material will be highly dependent on the level and consistency of contamination in the waste.

The final step in the waste reduction study is the financial analysis. After engineering and technological solutions have been found, equipment must be selected and sized, and costs developed. This is the area for the accounting representative on the team. All too often, engineers and process personnel gloss over the financial aspects of projects. If waste reduction projects are going to succeed, they must succeed on a cost-savings basis. Therefore, the financial department should be involved as early as possible; if they develop a cost-savings analysis favoring the project, the project is nearly sold.

The final phase of this process is to implement the waste reduction recommendations. Implementation, with regard to equipment installation, is no different than any other engineering or mechanical project. Implementation of procedural changes may be more difficult and time-consuming.

Once the recommendations are implemented, the quantity of waste reduction and the overall savings achieved by the effort should be monitored and regularly evaluated. Evaluation and recordkeeping are essential to illustrate the benefits of the waste reduction effort.

The waste reduction assessment should be repeated periodically (say, every three to five years) to ensure that the minimum amount of waste is being produced and to check on other procedural changes affecting waste reduction efforts.

Waste Reduction Techniques

Source Reduction

Source reduction means the reduction or elimination of waste at the source before it is generated. This technique can apply to waste of all types: solid, liquid, or hazardous. The only truly effective method of controlling the escalating costs of managing wastes is to prevent their formation. This fact is evidenced by the EPA's recent creation of an Office of Pollution Prevention to give waste minimization a high priority and profile.[49] This approach is fundamentally different from past regulatory schemes, which were primarily concerned with controlling pollution that had already been generated. The change in focus is perhaps the inevitable result of recognition that many of the benefits of controlling pollution have already been achieved and that further environmental gains can come only from eliminating pollutants at the point of generation.

Those undertaking source reduction activities must take a broad perspective of manufacturing and production processes. To evaluate source reduction opportunities properly, each process or manufacturing operation must be closely examined to determine material inputs, transformations that occur as part of the production process, and material outputs. The effect of quality control parameters, product specifications, and production goals must also be considered.

Source reduction techniques are generally divided into four basic categories: good operating practices, technology changes, material changes, and product changes. Each of these techniques is discussed below.

Good Operating Practices

Good operating practices are procedural or administrative measures that can be used to minimize waste generation. These practices address the human element in production processes by identifying areas that require attention to prevent waste generation. Many of these measures typically have been used to improve efficiency and reduce production costs. Improving yields by reducing production losses has long been common practice in industries where raw materials account for a significant portion of operating costs. Good operating practices generally require little or no capital investment, are easily implemented, and result in significant savings. Good operating practices include the following:

Waste reduction programs
Management and personnel practices
Material handling and inventory practices
Loss prevention through preventative maintenance
Waste segregation
Cost accounting practices
Production scheduling

Waste reduction programs should be formalized to indicate to the employees that the program is endorsed by management. Management should also attempt to be a visible supporter of the reduction program by recognizing company progress in meeting program goals and by identifying employees or departments that make significant contributions to the success of the program. Incentives, bonuses, or special privileges, such as an assigned parking space, are all effective means of rewarding employee contributions. If production bonuses are used in a facility, a formula that contains a penalty for increased waste generation can be included in the payout calculation. Communication of program goals, incentives for good practices, and penalties for poor practices are vital elements of effective management participation in a successful waste reduction program.

Material handling and inventory practices include programs that attempt to reduce generation of waste due to mishandling or expiration of useful life. This area can be a fruitful source of waste reductions. For example, a large metropolitan transportation agency uses grease to lubricate rails during winter. The agency's practice was to purchase grease in 55-gallon drums and, each winter, to place a drum in the vicinity of each area requiring lubrication and use the grease as needed. Typically, when the drums were removed in the spring, roughly half of the contents of each drum had been used, and the remainder had to be disposed of as a hazardous waste. By changing to the purchase of grease in 30-gallon containers, the agency eliminated the generation of this waste altogether because the contents were completely used during the winter.

Loss prevention minimizes waste generation by increasing maintenance to reduce leaks from equipment and spills. This is yet another technique that requires no capital investment but only diligent attention to production equipment. A large communications equipment manufacturer used a nickel-plating process, which contributed a significant nickel content to the sludge produced in the firm's wastewater treatment plant. Production personnel noted that even during periods when no material was processed through the nickel tanks, the wastewater still contained nickel. The firm instituted an extensive maintenance program and eventually identified the nickel source as a leaking flange, which was located below a grating and dripped into a trench that conveyed wastewater to the treatment system. The flange was replaced, thereby reducing the level of a hazardous constituent in the sludge, as well as raw material losses.

Waste segregation can reduce the volume of hazardous wastes generated by preventing the mixing of hazardous and nonhazardous wastes. Generally, if nonhazardous wastes are mixed with hazardous wastes, the resulting mixture must be managed as a hazardous waste. Common practices such as combining solvents for disposal can actually increase disposal costs if, for instance, chlorinated and nonchlorinated solvents are mixed together. Waste segregation can also be effective at reducing treatment costs. For example, if a plating operation uses hexavalent chrome or cyanide plating in the production process, segregation of the associated rinsewaters will likely reduce costs because a portion of the amount of treatment chemicals required is volume related.

Cost accounting practices should also be examined to ensure that the costs of pollution control and residual disposal are properly allocated to the areas or departments that produce the wastes. Accurate allocation will alert production areas to their actual production costs. Centralized accounting for pollution control and waste disposal leaves individual production areas with no incentive to control these costs. This information also provides effective data that can be used to evaluate the progress of individual departments and identify areas for investigation.

Production scheduling, especially where batch processes are used, can also be an effective waste reduction technique. For example, if a solvent manufacturer schedules production to ensure that line flushes do not cross-contaminate products, the material can be sold as a higher grade product.

Technology Changes

Technology changes, which involve direct modifications to production processes or equipment, can be significant avenues for reduction or elimination of waste. These modifications can range from minor alterations, which can be implemented quickly and at low cost, to major process replacements requiring significant capital outlays. These types of modifications can include the following:

Changes in the production process
Equipment, layout, or piping changes
Use of automation
Changes in process operating conditions, such as:

   Flow rates
   Temperatures
   Pressures and
   Residence times

Technology changes can also have secondary benefits such as improvements in health and safety. For instance, an electrical component manufacturer used a freon vapor degreasing operation in the production of electrical connectors. The connectors were placed in wire baskets, and manually immersed and removed from the degreaser. The removal velocities experienced in this process disturbed the freon vapor layer, resulting in significant freon losses. An automatic winch was installed to control the immersion and removal rate, which greatly reduced the vapor losses. The decrease in vapor losses also improved the work atmosphere by reducing worker exposure to freon vapors.

Input Material Changes

Input material changes are effective waste reduction techniques because they can reduce or eliminate the introduction into the production process of substances that cause the generation of hazardous wastes. Input material changes include material purification or substitution. Implementing such changes requires active participation by production, quality control, and manufacturing personnel.

In many cases, a material change can also increase productivity because of a decrease in rejection rates. For example, solvents are frequently used to clean printed circuit boards, but aqueous-based alkaline cleaning systems can be an effective substitute for this purpose. This substitution can increase productivity by allowing maintenance of more intimate contact between the cleaning solution and the circuit board, resulting in a cleaner product, fewer rejects due to insufficient cleaning, and the elimination of a hazardous waste (the spent solvent).

Product Changes

Product changes can be initiated by a manufacturer to reduce the generation of hazardous waste resulting from the use of a product. Product changes can include product substitution, product conservation, or changes in product composition. These types of changes are generally the most difficult waste minimization techniques to implement because unsatisfactory market acceptance may result in reduced market share for the product. Quite naturally, manufacturers are reluctant to alter a successful product.

Nevertheless, several examples of product changes that have gained acceptance in the marketplace exist. The paint industry has developed water-based coatings that replace solvent-based coatings in many applications. Water-based coatings do not contain the flammable or toxic solvents typically used in solvent-based paints and, accordingly, unlike solvent-based paints, are not RCRA-regulated hazardous wastes when they are disposed. Eliminating the solvent from the paint formulation also allows application equipment to be cleaned without using solvents. Yet another benefit resulting from development of water-based coatings is a significant reduction in volatile organic emissions.

The electroplating industry has developed many plating solutions that eliminate the use of cyanide. Zinc chloride and acid zinc solutions have successfully replaced zinc cyanide solutions in many applications. Similarly, copper chloride and acid copper solutions can often successfully replace copper cyanide solutions. These substitutions not only reduce treatment costs by eliminating the oxidation chemicals necessary to treat cyanide solutions, but also diminish the long-term liability associated with the disposal of sludges containing cyanide. In addition, once cyanide is eliminated from electroplating sludges, these sludges may require no additional treatment to comply with RCRA's restrictions on land disposal of hazardous wastes.[50]

Recycling

Once it has been determined that a waste stream or a component of a waste stream cannot be reduced or eliminated through the source reduction methods outlined above, recycling presents the next best opportunity for waste reduction. The first step in recycling a waste is to identify how the waste can be recycled and who the end customer will be. For example, a spent solvent stream is a good candidate for distillation recovery of the solvent in a pure form, and the distillate bottoms may have value as a fuel. A metal-laden sludge may have a metal content high enough to allow metal recovery from the sludge, and packing material, such as corrugated cardboard, once cleaned of contamination (wood, plastic) can be baled and sold to paper manufacturers.

Once a potential outlet for the material is identified and the waste stream thoroughly analyzed, requirements can be set with the recycler regarding the quality of the resulting product. Often the recycler will demand a product cleaner than the current waste stream. In these cases, methods to remove contaminants from the waste must be investigated. Often, the necessary methods may be as simple as waste segregation. Other treatment technologies that may produce a cleaner waste stream are specific ion exchange, evaporative recovery, and specific adsorptive removal of contaminants. As mentioned earlier, high levels or fluctuations of contamination in a waste stream may cause the recycler to reject the product and the recycling effort to fail.

Waste streams also often require concentration before recycling. For example, if the metal content in a sludge is below the economic threshold for metal recovery, the waste may be segregated and thereby concentrated sufficiently to increase the metal content of the sludge and give it an economic value. Baling of corrugated cardboard is another, not obvious, example of waste concentration. Shipping uncompressed and unbaled corrugated cardboard is not economically feasible. A third example is provided by the many waste streams that are heavily "contaminated" with water. Dewatering the waste stream serves to concentrate the waste, allowing the waste to be recycled and reducing the volume of material for shipment, thus lowering overall costs.

Often the producer of the waste can itself be the customer for recyclable waste products. An example of this type of recycling is found in the metal finishing industries. Metal fabrication processes use coolants to cool and lubricate metal cutting operations. Equipment is available that can recover these coolants and return them directly to the machining operation.

One area of potential industrial waste reduction that is commonly overlooked is municipal solid waste from industrial facilities. Industrial waste reduction efforts generally focus on hazardous waste because it is more highly regulated. Yet significant waste reductions and cost savings can be realized in industrial trash. Factory waste reduction efforts should consider the entire solid waste spectrum.

In conclusion, the process of waste stream recycling can be reduced to four steps: identify a waste stream's potential use; identify a recycler or end-use customer; segregate and concentrate the waste stream to increase its recycling value; and control contamination.

Case Studies in Waste Reduction

Case 1: Automotive and Truck Manufacture

The manufacture of trucks and cars involves the formation of various parts, machining of the parts, assembly, and painting. The manufacturing process is often divided into two parts—manufacture and assembly; therefore, this case study in waste reduction will also be divided into similar parts: Automotive Parts Production and Automotive Assembly.

Automotive Parts Production

The production of automobiles and trucks involves the production of parts from steel and other raw materials. The parts are then cleaned, coated, and assembled. The process of parts production produces numerous waste streams. In this case, the manufacturer had installed a centralized wastewater treatment plant to handle many of its waste streams. From a waste reduction perspective, a system that would segregate these streams so that each could be minimized would be preferred to mixing the streams and using an end-of-pipe treatment method.

A waste reduction audit performed at this facility identified many potential areas for waste reduction. In machining raw materials into the various parts, the manufacturer uses coolants and oils to assist in the cutting and drilling of metals. These coolants and oils were recirculated until they became contaminated and the emulsion began to break down. They were then discarded into the facility's sewer system. A cost savings of $19,000 a year was realized by segregating the oils from the coolants and instituting a coolant management plan, employing full recycling of the coolants through commercially available coolant recycling units.

The oil-coating process also created waste in the form of waste greases. Specific parts were dipped into a grease/solvent mixture and the solvent was then removed by heat treatment. During this entire process, grease dripped off the parts into collection pans. The low viscosity grease was recycled, but the high viscosity grease (primarily from the heating process), which required manual removal, was drummed and incinerated. The waste reduction audit revealed that if this high viscosity material were mixed with solvent, it could also be recycled to the process, producing a cost savings of $25,000 each year. In addition, any of the grease that could not be recycled, due to thermal breakdown or foreign body contamination, was suitable for use as a supplemental fuel. Sending this waste to a fuels blending program, rather than to a hazardous waste incinerator, yielded an additional savings of $6,000 each year.

The painting process in any production facility produces significant, difficult-to-handle waste streams. In this facility, parts were painted in a water curtain spray booth. The overspray from the air spray guns was captured in the flowing water. The paint/water mix flowed to a tank, where the paint either floated on the surface or coagulated and fell to the bottom. This paint sludge was removed and sent off-site for disposal. Several possibilities for waste reduction were presented by this process. First, the transfer efficiency of the spray equipment could be increased, thus increasing the amount of material on the final product and decreasing the amount of overspray in the water. Dewatering the paint sludge presented the potential to decrease the total amount of material requiring shipment for disposal and therefore, to decrease the cost of both transportation and disposal. In this particular case, the facility was reviewing their overall painting process and there was a possibility that the painting process could be eliminated entirely, thus eliminating these wastes.

The plating, phosphate, and miscellaneous wastes from the parts production process were sent to the centralized wastewater treatment facility. The wastewater treatment process employed lime precipitation of the wastes and produced a metal hydroxide sludge. The waste reduction audit revealed that too much lime was being employed in the process, resulting in excess sludge production. Installing process monitoring equipment and improving the lime addition system corrected this problem. Cost savings are being monitored with the new operation.

In the painting process, mask fixtures were used to create detail on many of the parts. When paint accumulated on the mask fixtures, they were stripped by immersion in methylene chloride. The methylene chloride was recycled through a still at the facility. The still bottom residues were drummed and shipped off for disposal. Methylene chloride losses in this system were significant, amounting to 300,000 pounds a year. Here, the solution was to change the process by applying a barrier coating to the masking fixtures. The barrier coating allowed the paint to be removed with a water spray. The water/paint mixture was then added to the spray booth wastes discussed above. Although this change did not reduce the amount of waste requiring disposal, it reduced air emissions and significantly reduced costs by eliminating approximately $62,000 each year in methylene chloride purchases.

In the process of assembling finished parts, many smaller components were purchased. These components arrived at the facility in a variety of packagings, all of which became solid waste. It was estimated that 60 percent of this waste was corrugated cardboard, which could be recycled. In addition, some of the specialty packaging (e.g., styrofoam holders) could be reused to package the final product. Recyclable plastics from the production line were also discarded. The facility had an office paper recycling program in place, but recycled volumes had steadily decreased and contamination had steadily increased. Therefore, the volume and value of this program had dropped significantly. It was estimated that the facility could save an estimated $50,000 a year by properly recycling cardboard, plastics, and office paper.

The overall identified savings from the waste reduction audit of the automotive parts production process was estimated to be $162,000 a year and the waste output from the facility was reduced by 5,760,000 pounds.

Automotive Assembly

The automobile and truck assembly process is similar to the parts production process, but it is on a larger scale. Automobiles are built from a large number of prefabricated parts. These parts arrive at the assembly plant in a wide variety of packaging materials. As parts are used, this packaging builds up on the assembly floor and is periodically moved to a compacting area. In the facilities audited, all of the packaging was compacted with other solid waste and removed for landfill disposal. This solid waste contained significant amounts of cardboard, clean plastic, and wood pallets. Corrugated compaction facilities were in operation. Nevertheless, due to financial incentives in the disposal contract, it was more lucrative for the disposal company to landfill the cardboard than to recycle it.

Waste reduction recommendations included renegotiation of the solid waste disposal contract to give the disposal company possession of the corrugated bales and the potential revenue from their sale. In addition, in the renegotiated contract, waste disposal costs were based on weight rather than volume. This change had the effect of increasing the compaction of the solid waste significantly so that the waste company could maximize the weight for each load. The total effect of these solid waste handling changes amounted to $700,000 each year.

Many of the automobile interior parts are covered in plastic to avoid soiling during transportation. Much of this plastic is removed before assembly, a waste of high-grade, low density polyethylene. Existing operations included hand compaction of these plastic bags into one bag. When the bag was full, it was thrown into a trash container. This waste stream accounted for nearly sixty tons each year of solid waste. The recommendation was made to further compact this plastic, bale it, and market the plastic to recyclers. With the revenue generated and the disposal savings, the company gained $40,000 every year from this single change.

The consultants also proposed employee incentive programs that dedicated a percentage of the revenue or savings from recycling to an extra employee benefit. This program helped keep employees involved and reduced contamination.

Many adhesives, sealers, and coating products, which come in drums, are used in the assembly of automobiles. In this study, it was found that the products were not fully removed from drums and that drums were not labeled or segregated. The result was an unworkable drum management system that allowed empty and partially full drums to accumulate. All of the drums were then disposed as hazardous waste, regardless of the material they had contained. Waste reduction recommendations included ensuring that drums were empty before they were changed on the assembly line; labeling empty drums with their past contents; and segregating drums that had contained nonhazardous materials from those that had held hazardous materials in the drum disposal area. An on-site drum crusher was also recommended to reduce the volume of empty drums leaving the facility. Finally, for products that came in open topped drums, the consultants recommended that the supplier add removable liners to the drums, allowing perfectly clean drums to be returned to the supplier for reuse. Savings from improved drum handling were estimated at $75,000 a year.

The final step in the production of an automobile is the application of a high quality finish coat. The industry standard is a paint coating applied by air spray equipment. Although improvements have been made to increase the transfer efficiency of the painting process, overspray still occurs. This overspray is captured by a falling film of water along the spray booth walls. The water flows to basins, where the paint is detackified by adding chemicals. The paint falls to the bottom of the tank and accumulates as a sludge. The water is then returned to the spray booth. Significant potential savings were identified in connection with the disposal of this sludge. A 44 percent reduction in volume and a savings of $40,000 a year were made possible by dewatering the sludge prior to disposal.

Overall identified waste reduction cost savings in the automotive assembly process were $855,000 each year and waste output was reduced by 34,600,000 pounds.

Case 2: Machining and Finishing Operations

The facility presented in this case manufactured components for defense related equipment. The primary operations in the facility were metal forming, machining, and metal finishing. When this facility was audited, three major waste streams were identified as candidates for waste reduction.

In the machining process, metal is formed, cut, and drilled to produce various parts needed for defense products. Oil and coolants are used in the process to extend the life of the cutting tools. With time, these coolants became contaminated with metal filings and were routinely discarded when it appeared that they were no longer performing. Disposal of the coolants cost approximately $105,000 a year. By instituting a coolant management program using full recycling of the coolant, it was estimated that the facility could save $200,000 each year in disposal costs and the cost of coolant purchases.

The finishing process used in this facility was an anodizing process. In the process, the metal parts were immersed in several solutions and rinses. Each solution and rinse bath had specified operating ranges. Once a solution exceeded the range, the contents were poured into a drum and sent for disposal.

Waste reduction recommendations included neutralization of the acid solutions. By neutralizing the spent acidic baths, these wastes were no longer hazardous and could be discharged to the sewer. RCRA regulations currently exempt elementary neutralization from permitting requirements when neutralization is the only treatment undertaken. Therefore, treating these wastes in this manner did not require a RCRA permit. Solids that occur in the neutralization process must be removed prior to discharge, however. Neutralization was projected to save $45,000 each year in disposal costs.

Chromic acid and sodium dichromate were used in the anodizing process. The spent solutions from these tanks were hazardous because of their pH and chromium content. The existing practice was to drum these wastes and dispose of them without treatment. The waste reduction consultants recommended installation of a closed loop recovery system, which consisted of cation and anion exchange filters, an electrodialytic unit, and ancillary equipment. This system recovered chromic acid for reuse, purified the rinse waters, and reduced the volume of waste requiring disposal by 95 percent. This system had an associated cost savings of $130,000 annually.

The third largest waste stream at this facility were used 55-gallon drums. The facility was spending $72,000 each year to dispose of these empty drums. When the vendors of the various products used at the facility were contacted, it was discovered that several sold their products in returnable drums or other returnable containers. Others were willing to work with the facility by adding liners to the drums, which would allow the drums to be returned and reused. Switching to returnable containers and working with the product vendors to return empty drums reduced the number of drums requiring disposal by 38 percent.

In connection with the drums, the waste reduction audit also disclosed that the purchasing department did not account for the cost of nonreturnable packaging when evaluating lowest cost purchasing. The consultants recommended that this prepurchase evaluation should add the cost of drum disposal ($20) to the cost of each drum of material sold in nonreturnable drums. This accounting change would encourage purchasing material in returnable containers.

In another area of the facility, the consultants found that reconditioned drums were being purchased for waste disposal. A plan to mark empty product drums as to their acceptable use for waste disposal was developed. This management change seems very obvious, but a waste reduction audit was needed to bring these two departments together to facilitate drum reuse. For the remaining drums, it was recommended that the facility purchase a drum crusher to reduce the volume. Overall, the recommended changes in drum management saved the facility $60,000 each year.

The overall cost savings achieved under Case 2 was $435,000 annually and waste output from the facility was reduced by 1,100,000 pounds.

Case 3: Waste Reduction Examples from Dow Chemical[51]

The agricultural chemical business has achieved significant reductions in contaminated container volumes by working with contract packagers. DURSBAN 50W, a wetable powder insecticide widely used in the landscape maintenance and horticulture business, used to be sold in 2-pound metal cans that required decontamination prior to disposal. This water-soluble powder is very dusty and posed a significant potential for exposure by inhalation and skin contact if not used properly. Dow now packages this product in 4-ounce, water-soluble packages, which can be disposed like any household waste.

Another example of successful waste reduction arose in the shipment of an activated ingredient for use in an insecticide formulation. This ingredient was formerly shipped in 55-gallon metal drums that had to be decontaminated and crushed before disposal. The product is now shipped in tank cars. When the tank cars require cleaning, they are decontaminated with a solvent used in the formulation of the activated ingredient.

One production facility investigated ways to reduce the amount of effluent leaving a crude product drying system. The drying agent is a purchased item, and the production staff saw the chance to reduce both potential environmental problems and costs. Initially, operations people added the drying agent manually based on production with a standard minimum flow. Samples were run six times a day with an average sample volume of 3 gallons a day. The engineering and production staff then developed a system that allowed the computer to ratio the drying agent addition based on feed flow (feed forward) and an on-stream analyzer was added to replace lab analysis and reset the ratio to a minimum (feed back). This project yielded immediate results, both in waste and cost reduction. A measured reduction of 37 percent in effluent volume and a corresponding reduction in purchased material cost were realized. The return on investment for this project, including manpower, was 67 percent.

Yet another example of a successful waste reduction measure is the use of process condition changes and recycling in a chlorinated hydrocarbon process to reduce waste volume and increase profitability. This reduction was accomplished in three phases over a two-and-a-half-year period. Phase I involved changing the operating temperature on condensing column bottoms. The subsequent shift in solubility of chlorinated hydrocarbons resulted in a 30-percent reduction in the chlorinated hydrocarbon volume in the effluent. In Phase II, the addition of a residence time reactor ("bump in line") and heat further reduced chlorinated hydrocarbon volume by 90 percent. In Phase III, the addition of a post-reactor stripper reduced chlorinated hydrocarbon volume to 99.9 percent of Phase II levels. The quality of the effluent was suitable for use in other production facilities. The plant realized a 2-percent carbon yield increase for all three phases and a 99.999-percent chlorinated hydrocarbon reduction in effluent. The overall project return on investment was 40 percent.

The final example demonstrates that waste reduction can be achieved through an investment in personnel. Application of simple statistical tools (Pareto charts, histograms, X and R charts) identified and solved problems within a production process and reduced production of off-specification product, which required treatment before disposal, by 4 million pounds a year. Here an investment in personnel yielded additional profits of $100,000 annually.

TABLE 2

SOURCES OF MATERIAL
BALANCE INFORMATION

- Samples, analyses, and flow measurements of feed stocks, products, and waste streams

- Raw material purchase records

- Material inventories

- Emission inventories

- Equipment cleaning and validation procedures

- Batch make-up records

- Product specifications

- Design material balances

- Production records

- Operating logs

- Standard operating procedures and operating manuals

- Waste manifests

TABLE 3

FACILITY INFORMATION FOR
WASTE REDUCTION ASSESSMENTS

Design Information

- Process flow diagrams

- Material and heat balances (both design balances and actual balances) for

    - production processes

    - pollution control processes

- Operating manuals and process descriptions

- Equipment lists

- Equipment specifications and data sheets

- Piping and elevation plans

- Equipment layouts and work flow diagrams

- Environmental information

- Hazardous waste manifests

- Emission inventories

- Biennial hazardous waste reports

- Waste analyses

- Environmental audit reports

- Permits and/or permit applications

Raw Material/Production Information

- Product composition and batch information

- Material application diagrams

- Material safety data sheets

- Product and raw material inventory records

- Operator data logs

- Operating procedures

- Production schedules

Economic Information

- Waste treatment and disposal costs

- Product, utility, and raw material costs

- Operating and maintenance costs

- Departmental cost accounting reports

Other Information

- Company environmental policy statements

- Standard procedures

- Organization charts

MUNICIPAL SOLID WASTE

As with industrial waste, the goal of reducing municipal solid waste (MSW) can be achieved through either source reduction or recycling. Each of these means is discussed below.

Municipal Source Reduction

Source reduction has great potential for reducing MSW over the long term. Over the last thirty years, however, the United States has experienced a steady upward trend in per capita waste generation: from 2.65 pounds for each person each day in 1960; to 3.22 pounds in 1970; to 4.0 pounds in 1988.[52] Source augmentation, rather than reduction, has been the rule. A further increase to 3.94 pounds for each person is projected for the year 2000.[53] If source reduction of MSW is to be achieved, manufacturers and consumers must change their practices—either spontaneously or in response to governmental policy initiatives such as those discussed in Chapter IV.

In examining source reduction of MSW, one encounters two approaches: (1) reducing the amount of waste generated and (2) reducing the level of toxicity in the waste generated. Both of these approaches are vital to the achievement of the economic and environmental goals driving the push for waste reduction.

Reduction in Amount

Reduction in volume (and weight) is the side of source reduction that addresses the growing shortages in capacity at waste management facilities. A number of factors affect the ability to achieve reductions in MSW production, including society's increased use of disposable goods, product longevity, ease of repair, compactness and economy of size, packaging trends, and process changes. The development of degradable plastics, although conceived by some as a volume reduction measure, is of questionable value for this purpose.

Disposable Goods

The development and growth in popularity of disposable goods has dramatically augmented the waste stream over the last fifty years. Many items that were once used repeatedly are now used once, or only a few times, and discarded. Consider disposable razors (which bear cadmium-containing orange pigments), disposable flashlights, paper towels and napkins, diapers,[54] syringes, and a host of other items.

These items did not emerge without reason. Some cost less than reusable items. Others offer convenience or protection against health risks. Disposable diapers are a boon for harried parents, although diaper services provide another, less waste-intensive alternative to washing cloth diapers. Disposable syringes avoid the risk of contamination and, thus, provide comfort to nervous patients in the era of AIDS. Whatever the motive behind these switches in consumer preference, the result is an ever increasing flow of MSW.

Longevity

Longer-lasting goods produce less waste. Thus, some technological advances counteract the increases caused by such trends as the development and increased use of disposable goods. Since 1983, for example, the useful life of dry cell batteries has increased by 45 percent.[55] Currently, manufactured car tires last one-half the life of a car. Two decades ago, tires were far less durable. Tire production by domestic tire manufacturers has dropped dramatically because of this change, and significant disruption has occurred within the industry. Nevertheless, there is a positive result—fewer tires, which are bulky and difficult to destroy, enter the MSW stream.

Ease of Repair

In the 1960s, fixing a dented fender on a car involved sanding the dent, putting in filler, and painting the wound. Now, the fenders on most cars are plastic. When dented, they are replaced and the old pieces are thrown away. Many electronic devices, including watches, are now less expensive to replace than to repair. The same kind of phenomenon has affected other items. This change has its virtues, such as speed of replacement and quality of repair, but, again, it results in the production of more waste.

Compactness

A positive trend for waste reduction efforts is that a number of items today are made more compactly than in the past. Small cars give rise to less waste than large cars. The United States has seen a shift to smaller cars, although that change is related to manufacturing and consumer energy costs, rather than concern with disposal. Similarly, less paper is used and thrown away when printing is on both sides of a page. Less container waste is produced if a box of rice or cereal is sold full—without an airspace at the top comprising a good share of the box's volume. If function is not affected, compactness will reduce waste volume and, therefore, should be encouraged.

Economy of Size

The other side of compactness is economy of size. Use of a very large vehicle—e.g., a bus or train—will produce less waste than the alternative use of many cars, small or large. The same effect occurs when people do other things jointly. Cafeteria meals, for example, produce less waste than meals prepared in single residences for the same number of people. The cafeteria will use big food containers in place of many small ones, and there will be less packaging material thrown away for each meal served.

Packaging

Containers and packaging made up 34 percent of gross MSW discards in 1986 and, whether or not deserved, clearly hold the status of public enemy number one in legislative initiatives to reduce waste.[56] Technological innovations produce new forms of containers. For example, the current popularity of microwaveable meals has produced a whole new spectrum of packages on grocery store shelves. Packaging and containers have functionally important design aspects, such as preventing breakage or retaining heat in food, that may not lend themselves to change for the purpose of waste reduction. Some commodities have more packaging, however, than is apparently necessary (e.g., powdered soap in packets, which in turn are in a plastic bag, which in turn is in a box). To the extent consumer awareness of the problems caused by such excess packaging can be increased, change may be accomplished.

Use of Less Wasteful Processes

As in the industrial waste field, process changes can have a positive effect on MSW generation. For example, transmitting electronic files produces less waste paper than sending printed documents in the mail. Similarly, advances in resin technology have reduced the amount of material in plastic bags and plastic milk jugs.[57]

Degradable Plastics

The much discussed issue of degradable plastics is more complicated than is generally appreciated. Bio- or photodegradable plastic in littered waste serves the purpose of making the waste's unattractive presence impermanent. If litter ends up in the sea (or the Great Lakes), its degradability will also help reduce the mortality of marine life that can result from entanglement. Degradation is of debatable value, however, if plastic wastes are properly managed.

Even plastics that are designed to break down degrade very slowly in a landfill. This unfortunate fact also holds true for paper, which can persist for decades or even centuries in a landfill.[58] At its most rapid pace, degradation of plastics in a landfill is not likely to prolong landfill life significantly through volume reduction. In fact, if the volume of trash in a landfill were to decrease substantially after closure because of degradation, any resulting subsidence in the landfill surface would likely disrupt a planned end use (e.g., a golf course) or closure systems.

An additional concern prompted by degradable plastics is the potential for leachate contamination by products of incomplete degradation. Some have argued that these degradation products will help produce a buffered "soup," which will reduce potential contamination, and that the degradation products may be inherently safe. Little hard information is available, however, one way or the other. Without hard information, all in all, something is to be said for dry, inert deposits in landfills.

In short, use of degradable plastics is not necessarily a significant positive step in MSW reduction.

Reduction in Toxicity

Heavy metals found in pigments and batteries warrant priority in minimizing toxicity in MSW. These products have been a significant source of metal contamination in the past.

Efforts to reduce their presence in the MSW stream have met with some success. The Eveready Battery Company, for example, has prepared a graph of its mercury usage showing a decrease in tonnage from 300 tons in 1983 to less than 100 tons in 1988.[59] The company's battery sales apparently did not decrease in the same period.

Three classes of inorganic pigments—metal oxides, lead chromates, and cadmium—introduce lead, cadmium, nickel, mercury, and chromium to the MSW stream. A relatively small number of companies make these pigments.

Metal oxides are used in very expensive, industrial coatings. They color and protect exterior construction projects, coils, wires, vinyl siding, auto parts, ceramics, and, in some cases, tiles and tableware.[60]

Lead chromate colors are used extensively in the manufacture of printing inks and coatings. Many consumer products that eventually become waste contain these pigments, including packaging materials, posters, gift wrap, and publications.[61]

Cadmium pigments are used in more than 50 percent of all colored plastics. They provide bright, light-fast yellows, oranges, and reds that can withstand extreme heat in processing. Products colored with cadmium pigments are in common use and are prevalent in the waste stream. These include personal care products and packaging, recreational equipment, and office equipment.[62]

Presumably, alternatives exist for many current uses of heavy metals in pigments. Most newspapers have already discontinued use of pigments containing heavy metals. The American Newspaper Publishers Association (ANPA) has adopted a list of industry standards for colored and black inks, which excludes pigments containing lead, chromium, or cadmium.[63]

Heavy metals in pigments and batteries are obvious "hot spots" in the search for means to reduce toxicity in MSW. Other special problems probably exist and may in turn receive the close scrutiny that has recently fallen on pigments and batteries. Higher volume discards with lower levels of toxic compounds also warrant attention. Policy initiatives discussed in Chapter IV would establish mechanisms for continuing review of potential toxic contributions to MSW.[64]

Municipal Recycling

Source separation, the process of sorting certain wastes for recycling, is an essential ingredient of successful recycling programs. Separation may occur at the residential household or at the commercial business. The importance of source separation is best conveyed by the description of various types of recycling programs now in use.

This overview discusses the most important postconsumer recycling programs in the United States:

Curbside residential recycling
Multifamily residential recycling
Commercial and industrial recycling
Office paper recycling
Yard waste collection and composting
Tire recycling

The remainder of the section is devoted to a discussion of material recovery facilities and mixed-waste sorting technologies that facilitate recycling efforts. This chapter does not repeat the discussion of market issues with regard to recyclables that appears in Chapters II and IV. The reader is reminded, however, that even the best program for collection of a recyclable item will fail if no market exists for the material. If no one wants the old newspapers that are diligently separated and placed out on the curb for collection, a homeowner may just be throwing them away anyway.

Curbside Residential Recycling Programs

Curbside collection is the most effective and popular option for single-family dwelling residential recycling because of its convenience. Residents set out specified recyclable materials on days designated for collection. Collection may be at the curb or from the backyard or alley. Because most residential programs collect from the curb, the term "curbside collection" has been coined. Recycling collectors pick up the materials and deliver them directly to markets or to their own processing and storage sites for later shipment.

Sonoma County, California, is one of many examples of successful residential recycling programs. Residential recycling began in 1978, providing collection and processing services to more than 52,000 households.[65] Service is provided for duplexes in addition to single-family homes. Recyclables are collected weekly on the same day as regular garbage collection. Each household is provided with three plastic bins for (1) newspaper, (2) glass, and (3) steel and aluminum cans and plastic soft drink bottles (PET) and milk jugs (HDPE). Participation by 55 percent of the households is producing 400 tons a month of recyclable materials.

Many communities are requesting proposals for curbside collection of recyclables as part of their solid waste management programs. Curbside collection is the most visible type of recycling program. Its effectiveness depends on participation by individual citizens. The level of participation obtained by these programs is influenced by a number of factors, including the nature of participation, frequency of collection, provision of household containers, and publicity.

Mandatory versus Voluntary Programs

Mandatory recycling programs generally have somewhat higher participation rates than voluntary programs even though the majority of communities that have mandatory separation ordinances do not actively enforce them. The key to wide participation, however, is convenience. If the importance of this factor is kept in mind, voluntary curbside programs may achieve 60 percent or higher participation. *See* **Tables 4** and **5** at p. 50 *infra*.

Frequency of Collection

Most programs collect recyclables weekly because a weekly collection attracts greater participation and diverts more materials to recycling than less frequent collection. One problem with infrequent collection is that people tend to throw their recyclables away when storage space runs out. Another problem is that people apparently will save their recyclables only during the week they will be collected, and throw them into the waste stream during the other weeks. Thus, weekly collection results in higher diversion rates and lower costs per diverted ton than biweekly or monthly collection. *See* **Table 4** at p. 50 *infra*.

Collecting recyclables on the same day as regular garbage service yields a 5 to 10 percent higher participation than collection on days other than regularly scheduled garbage collection. Same-day recycling collection is apparently more effective because those citizens participating are better able to recall when they are supposed to set out their recyclable materials.

Provision of Household Containers

By providing household containers for the accumulation of recyclables, a program can double—or even quadruple—participation rates. *See* **Table 5** at p. 50 *infra*. These containers provide a daily reminder to recycle. Placed at the curb, distinctively colored and labeled, the containers also increase public awareness and exert peer pressure on neighbors.

Publicity

The effort devoted to promoting a recycling program has a marked effect on participation. Communities with intense publicity campaigns report an immediate positive impact on citizen behavior. For example, program officials in Palo Alto, California, noted a significant increase in use of their curbside program when their recycling vehicle was featured on the front cover of the municipal telephone directory.

Providing recycling information through flyers or door hangers has proven particularly effective in stimulating participation. Educating residents not only helps recycling programs, but, in the long run,

helps the community by saving landfill space and preserving the
environment.

TABLE 4
EFFECT OF COLLECTION FREQUENCY
ON PARTICIPATION[66]

Voluntary Programs	Collection Frequency	
	Weekly	Monthly
Number surveyed	17	14
Participation range	10%-80%	4%-65%
Average participation	46%	29%
Mandatory Programs		
Number surveyed	9	6
Participation range	40%-98%	25%-85%
Average participation	73%	48%

TABLE 5
EFFECT OF HOUSEHOLD CONTAINER
USE ON PARTICIPATION[67]

Test Community*	Program Participation in %	
	With Container	Without Container
Champaign, IL	83	11
Kitchener, Ontario	75	65
San Jose, CA	75	48
Santa Rosa, CA	70	35
Toronto, Ontario	66	42

* Programs are voluntary

Multifamily Residential Recycling

Recycling from multifamily dwellings is an important aspect of residential recycling due to the large number of people in such housing and the high tonnage of MSW they produce. The term "multifamily" refers to a broad range of dwellings, including retirement homes and dormitories, but recycling services to date have focused primarily on apartments, condominiums, and co-op housing units. As for single-family dwellings, newspapers, glass bottles, aluminum and steel cans, and plastic bottles are the principal recyclables collected.

In many high-rise apartment buildings, residents deposit their refuse down chutes that feed into compactors or containers, which are periodically removed by municipal or private haulers. Recycling operators often place containers inside or immediately outside these buildings; tenants can then deposit their separated materials into these containers for recycling. Some high-rise buildings do not allow newspapers and glass containers to enter the trash chutes because of potential clogging problems. Residents in these buildings generally stack their newspapers and glass bottles on the floor in the regular refuse room. The building maintenance staff then separates the material for recycling and transfers it to depot containers for pickup. A curbside approach, in which tenants set recyclable materials out at the curb for separate collection in the same manner as single-family dwelling residents, is used only in very small apartment buildings (often of four units or less).

In general, a depot approach, modified to suit particular circumstances, is used at low-rise apartments. Size and placement of the bins vary according to building size and the space available, both inside and out. Some recycling operators place stackable bins inside the complex to help people pile and tie their newspapers, or barrels to collect glass bottles and cans. Others merely assist the building manager in setting aside appropriate space to collect the materials. Where indoor space is not available or not acceptable due to fire code regulations, materials can be stored outside the building in a centrally located place, generally the complex parking lot.

Recycling operators will generally not service places to which they cannot drive directly (i.e. underground parking garages or walk-ways). Pickup frequency varies depending on the volume of material and location of the complex. At some units, the operator may wait until the building superintendent calls with a report that the bin is full. At others, pickup varies from twice weekly to twice monthly and may correlate with the regular garbage pickup day. The containers are either emptied on site or removed and replaced with new ones. The materials are then transported to a buyer or stored before processing.

The city of Robbinsdale, Minnesota, for example, provides apartment buildings with recycling services. Residents may place their recyclables in separate dumpsters that are provided by the city. Waste Management of Blaine collects and processes the materials every two or three weeks, or more often if necessary. In a pilot program run at eight buildings with 437 units, each unit averaged 3.3 pounds of recyclables a week, of which 82 percent was newspaper.

Commercial and Industrial Recycling

Recycling of MSW from commercial and industrial sources works well when employing hauling companies, transfer stations, and landfills. Transfer stations can often be modified to act as intermediate processing centers for recyclable materials. A portion of the transfer station can be used to process materials such as corrugated cardboard, paper, and pallets. Doing so usually produces a more cost-efficient operation, better utilization of available space, and improved community relations.

Recycling can also be incorporated into a landfill development and operation strategy. Areas separate from the active face of the landfill are most suited for recycling activities, such as composting, wood and tire reclamation, and demolition waste processing.

In the past, commercial and industrial recycling programs were thought to be operations that collect and process only old corrugated cardboard. Today, materials such as glass, foundry sands, wood, office paper, and construction materials are being collected and recycled from commercial accounts. There is less public awareness of the significant recycling potential presented by these materials than the familiar items associated with curbside programs. The following discussion highlights the types of material that can be recovered through a commercial/industrial recycling operation.

Aluminum

Aluminum has one of the highest market values of any material routinely collected in a recycling program. When aluminum scrap is recycled, 95 percent of the energy required to produce aluminum from ore is saved. Because of this impressive energy savings and other economic reasons, aluminum scrap has become an increasingly important source of supply for the industry.

Major sources of aluminum beverage cans are airports, bars, restaurants, hotels, and sporting complexes and stadiums. Sources of other types and alloys of aluminum are construction and demolition sites, storm window fabricators, mobile home manufacturers, and all other aluminum product fabricators.

Construction and Demolition Waste

There are seven major types of recyclable material that can be recovered on a large scale from construction, renovation, and demolition sites. These are steel, roofing materials, aluminum siding, asphalt, concrete, bricks, and wood. Other materials such as copper, piping, electrical fixtures, wood cabinets, and doors may also be recovered and resold, usually by demolition contractors.

Great quantities of steel waste are often produced when large buildings or factories are demolished. Recovery of this material is sometimes done by the contractor. Even if the market value of the material is low, recovery may present an economically viable alternative to landfilling in areas where disposal costs are based on a per ton basis.

Roofing material wastes contain a mixture of fibers, aggregate minerals, and petroleum-based materials. The majority of these wastes are produced during the warmer months of the year and in the older sections of cities or towns. Material that is separated and recovered can be incorporated into asphalt materials used in the construction of roads. The market value of this material is low to nonexistent, but, as with steel wastes, the avoided or reduced cost of disposal in areas with high disposal costs sometimes makes recycling roofing material an attractive option to roofing contractors.

Aluminum siding, which over the past couple of decades has become a popular substitute for painted wood exteriors on homes, is an increasingly available source of recyclable aluminum. The amount of waste produced when aluminum siding is installed on a home is relatively small. When homes having aluminum siding exteriors are renovated or demolished, however, the potential to recover and recycle the aluminum siding is presented.

Asphalt is often recycled to make new asphalt. This practice may reduce transportation costs and often eliminates associated costs to dispose of the material. The recyclability of asphalt is limited by the exact nature of the material to be recycled, the end use of the material, and most importantly, the presence of a local asphalt recycler.

Concrete is commonly recycled to make aggregate material, incorporated with cement to make new concrete, used as base material for roads, or used to construct a drainage layer in landfills.

Bricks that are undamaged and in relatively good condition can be recovered from demolition sites. Although the recovery is very labor intensive, certain types of brick and stone can offer a relatively high resale value, if an end market can be found.

Wood waste may constitute up to 80 percent of the total waste generated at construction and demolition sites. Sources of wood waste, however, are not limited to construction and demolition sites. Furniture manufacturers, warehousing and shipping firms, landscape companies, and municipalities also produce wood by-products that are generally discarded into the MSW stream. Once wood wastes are collected and chipped, using either a wood grinder or tire shredder, their end-uses include fuel, landscaping, and composting.

Food Wastes

Food wastes are generated in the processing of fruits, vegetables, meat, poultry, fish, and dairy products. Four hundred million tons of food wastes are generated annually in the United States. Some of this waste is composted and used for land improvement. Much of the waste that is recycled is used as feed for swine, poultry, fish, and pets.

Glass

With the exception of aluminum beverage containers, no other material has experienced such rapid growth in recycling as glass. Recently, there has been a growing effort to collect glass from commercial establishments such as bars, restaurants, hotels, multifamily dwellings, casinos, sports complexes, airports, and bowling alleys. These establishments produce large volumes of glass in a relatively short time.

Santa Clara, California, has joined the trend to collect glass for recycling from bars, restaurants, and hotels. Establishments producing 100 or more pounds of glass a month participate in the recycling program. Each week, collection vehicles pick up the glass, which is accumulated in separate containers supplied to the customer.

Paper

Paper and paper-related products are the most widely recovered materials in commercial and industrial recycling operations. The most common types of paper products recovered are corrugated cardboard, newspaper, and other grades of paper. At a major auto assembly plant, cardboard is collected, baled, and picked up by the mill on a twenty-four-hour basis. Three employees run the operation on each of three shifts. At this operation, cardboard recovery averages 250 to 300 tons a month.

Plastics

The goal of any recycling program using plastics from industrial accounts is to identify companies that produce a by-product or waste material in a relatively large, uncontaminated quantity. Industries suitable for targeting by a plastics recycling program are telephone manufacturers, auto parts fabricators, bottle and container manufacturers, computer companies, airports, packaging companies, warehouses, and auto manufacturers.

Office Paper Recycling

About 4 million tons of printing and writing paper are discarded from offices and other establishments every year. These papers are used in reports and letter writing, computer printouts, and many other applications. By far the largest volume comes from printed forms, which are frequently computer generated. However, banks, insurance companies, independent print shops, and some government office buildings are particularly large generators of waste printing and other office papers. Approximately 13 percent of all printing and other office paper is recycled by source separation.[68]

The corporate headquarters of Waste Management, Inc. in Oak Brook, Illinois, conducts such a source separation office paper recycling program. Participation in the recycling program is nearly 100 percent among 1,100 employees, and has reduced its waste stream by 50 percent. On average, 18 tons of paper—30 pounds a person—are collected each month. The national average in office recycling programs is 10 to 25 pounds a person.

The program incorporates simple procedures. A dual wastepaper basket system is used to collect white and colored ledger and computer paper. Each employee is responsible for emptying his or her container into one of numerous larger recycling bins centrally located in copy machine areas or coffee rooms. The cleaning crew then transfers the recyclable paper to larger bins located outside the buildings. The paper is stored there prior to pickup for transport to a paper processing facility, where the paper is graded and shipped to market.

The key to the success of this recycling program is planning. The company established a simple procedure for employees to follow, made clear to the employees that the project enjoyed the support of top management, and provided proper training of the employees and janitorial staff before implementation. Before the training session began, the company president sent a memo to all employees announcing the program and briefly explaining its procedures. After a brief training session, which included a slide presentation and discussion, the employees received their containers. Floor captains were designated on every floor of each building to monitor the program's progress and answer questions.

Yard Waste Collection and Composting

Yard waste represents an average of 18 percent of the municipal solid waste stream across North America. Depending on the region, seasonal peaks and lulls will greatly vary the amount of yard waste produced on a monthly basis. On a year-round basis averaged across North America, yard waste is composed by weight of 50 to 70 percent grass, 20 to 40 percent leaves, and 10 to 30 percent brush.

Traditionally, homeowners dealt with yard waste by burning leaves and brush and bagging grass clippings for regular garbage collection. This system was greatly altered in 1970 with the passage of the Clean Air Act. As a result, many incorporated communities implemented a ban on open fires and thus a ban to the traditional fall burning of leaves. Now, many state governments are looking toward banning yard wastes from landfills. Large scale, growing-season-long yard waste management programs are being implemented throughout the nation.

To implement a successful yard waste composting program, two key questions must be analyzed and answered: What are the potential end uses of the compost? How will yard wastes be separated and collected?

There are many uses for compost of varying quantities. Although the nutrient content of yard waste compost is too low for it to serve as a fertilizer, it is a valuable soil conditioner and organic amendment. It has the appearance of a fine, dark soil and forms a beneficial growing environment for plants. Compost provides improved soil texture and soil aeration, increased water holding capacity, decreased erosion, and improved regulation of soil temperature.

Yard waste can be collected with or without containers, or in plastic bags. Kirkland, Washington, is experimenting with the use of containers in its collection. This community of 560 households is divided into three sections. In one section, 90-gallon containers are provided for semi-automated collection of yard waste. A second area is provided with specially printed plastic bags, which are distributed at the customer's request. The third area is instructed to place yard waste in any rigid container. All three areas are instructed to place yard waste three feet from other garbage on their regular collection day. This program has successfully diverted 30 percent of the residential waste stream from the community's landfill.

Another successful program is in Albemarle, North Carolina, where residents place leaves and brush for collection in loose form. Leaves are collected with vacuums twice a month during the peak production months of the year. Brush is collected once a week all year round. Similarly, in Hastings, Minnesota, yard waste is collected during an eight-week period in the spring and in the fall. Residents can place their bagged yard waste along the curb for collection every other Saturday. In both the spring and fall, residents of the 5,000 households in Hastings divert 7 percent of their residential waste stream for composting.

Tire Recycling

American industry now draws only about 4 percent of its raw materials for rubber production from recycled sources.[69] Worn-out and rejected tires and inner tubes from passenger cars and trucks make up the bulk of scrap rubber that is available for recycling. Scrap tires can be recycled and converted to a soft, workable state suitable for use in the manufacture of products used in the automotive and agricultural industries. Recycled rubber is also used in the manufacture of asphalt-rubber materials used for roads, highways, sports and recreational surfacing, and airport runway construction.

Tires also can be used to make a good industrial fuel. In Franklin, Wisconsin, Waste Management, Inc. operates a shredder, which can process about 1,000 tires an hour, or one tire in two-and-a-half seconds. The shredder cuts the tires into small chips, which exit the machine on a conveyor and go into an open-top trailer. The tire chips are sold to a paper company, which mixes them with coal. The composite fuel is cleaner, more efficient, and more economical than coal alone.

Material Recovery Facilities

Material recovery facilities are used to sort and prepare recyclable materials for shipment to end use markets. Material recovery facilities process materials from a variety of sources within a community, primarily source-separated recyclable materials from residential curbside programs and commercial establishments.

Material recovery facilities are now often designed to receive, process, and market newspaper, corrugated cardboard, mixed paper, aluminum and ferrous metals, glass, and plastics. Source-separated materials are processed using conveyor systems and separation equipment. The equipment may include overhead magnets for ferrous removal and a series of air-classification and manual sorting techniques, which maximize processing efficiencies for plastics and glass recovery. After separation, materials other than glass are fed by conveyor into high-density balers and then shipped to market. Glass is crushed and shipped to end markets in containers.

Mixed Municipal Solid Waste Sorting Plants

Mixed waste sorting plants pull recyclables from unprocessed municipal solid waste. New mixed waste sorting technologies are emerging in the solid waste management industry. The technology operated by Waste Management, Inc., for example, is System Brini. The system is designed to recover fuel, compost, and ferrous metals from mixed household waste.

In the plant, the waste is first shredded into particles suitable for sorting. A ballistic classifier is then used to divide the waste into three fractions: light (paper and plastic); heavy (metal and heavy plastic); and compost (organic material). The light fraction normally consists of 70-percent paper and 10-percent plastic and is used as an alternative fuel. The heavy fraction is conveyed to a magnet, which removes magnetic particles. The ferrous waste is then baled for sale to markets. The compost fraction comprises organic waste, sand, and glass. It is often mixed with sludge before composting. System Brini may be used in conjunction with source separation at the curbside or elsewhere.

CHAPTER FOUR

Waste Reduction Policy

This chapter examines existing federal and state requirements and recent federal initiatives to encourage reduction of, first, industrial waste and, then, municipal solid waste.

INDUSTRIAL WASTE

Federal Regulations Affecting Industrial Waste Reduction Efforts

The federal government has as yet taken no direct action to require the reduction of hazardous waste. Federal certification and reporting requirements described below do encourage these efforts, however. In addition, federal technical assistance for waste reduction efforts is now available through the EPA's newly created Office of Pollution Prevention. This office maintains a national clearinghouse of technical information, including a computerized database of waste minimization information.[70]

Also discussed below are federal regulatory requirements applicable to recycling or reclamation of hazardous waste. These requirements vary a great deal depending upon the waste to be recycled and the particular form of recycling employed. Accordingly, identifying the applicable regulatory requirements is an essential step in assessing the feasibility of any hazardous waste recycling effort.

Before discussing these requirements, a brief overview of RCRA, the federal solid waste management statute, is in order.[71] RCRA establishes a comprehensive cradle-to-grave system for regulation of hazardous waste. The Act recognizes three classes of regulated parties: generators; transporters; and operators of treatment, storage, and disposal facilities. Any person who generates a solid waste is responsible for determining if the waste is a hazardous waste as defined by RCRA. If the material is a hazardous waste, the generator is responsible for notifying EPA that it is generating a hazardous waste. The generator is also responsible for ensuring that the waste is transported, treated, and disposed in accordance with the Act.

The EPA requires that every shipment of hazardous waste be accompanied by a manifest, which is signed by the generator and each person who subsequently transports, treats, or disposes of the waste. After the waste is disposed, the manifest is returned to the generator. The generator is responsible for ensuring that all manifests are returned to it and that the hazardous wastes were handled by the proper facilities.

A generator of hazardous waste may store wastes on site for up to ninety days without obtaining a permit. Treatment of the waste during this limited storage period does not trigger the requirement for a RCRA permit if the treatment takes place entirely within tanks or containers that meet RCRA specifications.[72] Anyone else who treats, stores, or disposes of hazardous waste must have a permit from the EPA.[73] Such facilities are subject to a broad range of EPA requirements, imposed both by regulation and as a condition of receiving a permit.

Certification and Reporting Requirements

Although RCRA does not impose any substantive waste reduction requirements on hazardous waste generators, it does impose several certification and reporting requirements. First, the generator must certify on each hazardous waste manifest that it "has a program in place to reduce the volume or quantity and toxicity of such waste to the degree determined by the generator to be economically practicable."[74] Second, in biannual reports required by RCRA, each generator must report on the efforts it has made to reduce the volume and toxicity of the hazardous wastes it generated and the actual reduction in volume and toxicity that has been achieved.[75] Finally, a generator that receives a permit to treat, store, or dispose of wastes generated on-site must annually make the same certification that is required on the manifest accompanying an off-site hazardous waste shipment.[76]

Regulation of Recycling Activities

RCRA covers only waste management activities; it does not apply to manufacturing processes. Because recycling and other types of waste reduction activities share aspects of both waste treatment and manufacturing, the status of recycled materials is somewhat ambiguous under RCRA. The EPA promulgated regulations in 1985 to define the regulatory status of recycled materials and recycling processes. These regulations are particularly arcane. The EPA's "definition of solid waste" regulations are perhaps the most complicated portion of the RCRA regulations. Their mere mention will elicit a groan from any lawyer familiar with the RCRA program. The authors therefore caution the reader to proceed very carefully and seek legal advice before venturing to rely upon a recycling exclusion or exemption from RCRA regulation.

In general, the EPA regulations attempt to distinguish between legitimate recycling and sham recycling (i.e., hazardous waste treatment activities posing as recycling to avoid regulatory requirements). The former is either exempt from regulation or regulated lightly. The latter is subject to the full range of hazardous waste regulation.

The regulations define three categories of legitimate recycling activities: recycling that is totally exempt from RCRA regulation; recycling that is subject to special regulation; and recycling that is partially exempt from RCRA regulation. To determine which category an activity fits into, one must look at both the material being recycled and the process being used. Note that the term *recycling* as used here and in the EPA's regulations is somewhat broader than the standard use of the term. As used here, recycling may include any process by which a waste or secondary material is beneficially used and includes the burning of waste for energy recovery, certain types of land application of waste, and the storage of waste in anticipation of future recycling.

Recycling Activities Excluded or Exempt from RCRA Regulation

The first category of recycling activities are those that are totally exempt from RCRA regulation. An activity can achieve this status in either of two ways: by exclusion from the definition of hazardous waste or by specific exemptions within the regulations. Certain types of materials, when treated in certain ways, are not, by definition, solid wastes. Because they are not solid wastes, they cannot be hazardous wastes under RCRA and, therefore, RCRA does not apply. Thus, activities or processes involving the recycling of these materials are not subject to the requirements of the RCRA hazardous waste regulatory program.

The regulations identify four classes of activities that are not solid waste disposal: (1) reclamation of sludges[77] and by-products[78] that have not been specifically listed by EPA as hazardous wastes and of commercial chemical products[79] (that are normally hazardous wastes when disposed); (2) materials that are used or reused as ingredients in an industrial process to make a product (provided the materials are not being reclaimed); (3) materials used or reused as effective substitutes for commercial products; and (4) materials that are returned as a raw material to the original process from which they are generated, without being reclaimed.[80]

To qualify as recycling under these exclusions, the recycling process need not be performed by the generator itself. However, if the generator relies on one of these exclusions to avoid classifying a material as a hazardous waste, the generator must be able to prove that there is a known market or disposition for the material. That is, if the generator claims that it does not need to manage a certain material as a hazardous waste because the material is going to be recycled, the generator must be able to identify specifically who will take the material and recycle it.

The definition of *reclamation* and *reclaiming* is obviously of crucial importance in determining whether a waste qualifies for one of these four exclusions. A material is considered to be reclaimed if it is either processed to recover a usable product (for example, the recovery of lead from spent batteries) or is regenerated (for example, the regeneration of spent solvents).[81] Essentially, if all of the waste is used, it is not reclaimed; if only part of the material is used, it is reclaimed.

Under the first exclusion, wastes from industrial processes or waste treatment processes can be reclaimed without regulation unless the wastes are specifically listed by the EPA as hazardous waste.[82] Additionally, commercial products that might otherwise be discarded (e.g., because they are off specification) can be reclaimed without being subject to regulation.

The second exclusion allows a waste material from one process to be used as a feed material in a second process to produce a new product, provided the waste is not reclaimed; i.e., all of the material must be used as is. The third exclusion allows a waste material to be used as a substitute for a commercial product (for example, the use of spent pickle liquor as a phosphorous precipitant and sludge conditioner in wastewater treatment or the use of hydrofluorosilic acid, an emission control dust, as a drinking water fluoridating agent). Note that if a new commercial use for a waste material is found, the waste would then be considered a coproduct and would not be subject to RCRA regulation. In either of these cases, the burden is on the generator and recycler to prove that the use of a waste is genuine recycling and not waste disposal. In distinguishing between sham and legitimate recycling, the EPA looks to whether a true commercial product is produced, whether the waste is actually used in making the product (as opposed, for example, merely to being contained in the product matrix), and the commercial viability of the recycling process on its own.

The fourth exclusion covers waste materials that are returned as feedstocks to the process in which they were produced as long as they are not first reclaimed. An example would be smelting wastes that are returned to the smelter. The regulations require that the waste material must substitute for raw material feedstock, but the process must primarily use raw materials for feedstock.

In addition to the four classes of excluded materials described above, there are a number of specific materials that the EPA has excluded from the definition of solid waste.[83] These include pulping liquors that are reclaimed in a specified manner; spent sulphuric acid used to produce virgin sulphuric acid; and secondary materials that are reclaimed and then returned to the original process as long as the reclamation and return process meets certain requirements.[84]

In addition to materials that are not regulated because they are excluded from the definition of hazardous waste, a number of recyclable materials are considered to be hazardous wastes but are nonetheless exempt from all regulation for generators, transporters, and recyclers.[85] These include industrial ethanol that is reclaimed; used batteries returned to a manufacturer for regeneration; scrap metal; fuels produced from refining of oil-bearing wastes; oil reclaimed from hazardous waste but generated from normal petroleum refining practices; certain coke and coal tar from the iron and steel industry; certain hazardous waste fuels produced from oil-bearing hazardous wastes; and petroleum coke produced from certain wastes.

Recycling Activities Subject to Special Regulation

The second category of recycling activities are subject to special regulation.[86] Each of these activities is subject to a particular set of regulations applicable to anyone who generates, transports, or recycles the waste.[87] The activities in this category are recycling materials by placing them on the land or using them to produce products that are placed on the land; burning of hazardous wastes for energy recovery in industrial furnaces and boilers; reclamation of precious metals from hazardous wastes; and reclamation of spent lead-acid batteries.

Recycling Activities Subject to Reduced Regulation

The third category includes all other hazardous wastes that are recycled. Generators and transporters of these hazardous wastes destined for recycling must comply with the same RCRA requirements that would be applicable if they were generating or transporting any other hazardous waste.[88] Moreover, an owner or operator of a recycling facility that stores this class of hazardous waste before recycling must obtain a storage permit and comply with all applicable storage regulations.[89] No permit is required, however, for the recycling process itself, and that process is not currently regulated.[90] The only requirements that now apply to a person who operates a recycling facility and does not store wastes before recycling are that the operator must notify the EPA of the facility's activities and comply with the RCRA manifest regulations.

State Hazardous-Waste Reduction Efforts

State governments have been much more active than the EPA in developing programs to encourage hazardous waste reduction. The first program was begun by New York in 1981 and, by 1988, at least thirty-six states had some sort of hazardous waste reduction program.[91] These programs vary from small-scale educational efforts to full-scale technical assistance programs. Although states undoubtedly can use their regulatory powers to mandate waste reduction, to date, very few have chosen this route. Most states instead have employed some combination of education and outreach, technical assistance, waste exchange, and economic incentive to encourage private hazardous waste reduction efforts. Several states have reported substantial success in reducing the amount of hazardous waste generated by industries in their states as a result of these efforts.

State Regulatory Programs

States have broad authority both to regulate recycling and other forms of waste reduction and to require waste reduction. Under RCRA, the EPA can authorize states to regulate hazardous waste activities within the state in lieu of federal regulation of such activities.[92] To receive federal authorization, a state RCRA program must be at least as stringent as the EPA's and may be more stringent.[93] Currently, more than forty states have received authorization.

In general, therefore, states are responsible for the certification and reporting requirements discussed above and may make these requirements more stringent than the federal requirements. They could require, for example, annual reports, rather than biennial reports, or more detailed information. States have generally viewed these requirements primarily as being intended to stimulate awareness of waste reduction, rather than as the basis for any regulatory activity.

States also have authority to alter the definitions of hazardous waste and recycled materials. In general, however, in this area, states have adopted the federal scheme described in the preceding section. States may also choose to impose additional regulations on recycling activities. The state of Louisiana, for example, has passed legislation requiring recyclers to comply with the same requirements as are applicable to hazardous waste treatment facilities.[94]

Finally, states can impose regulations mandating some form of waste reduction. So far, we believe the only state to have such a regulatory approach is Ohio. The Ohio program operates at the waste disposal level because a large number of disposal facilities are located in the state. Any generator (whether located in Ohio or not) that wants to send more than 200 tons of hazardous waste in one year to an Ohio hazardous waste landfill must obtain the state of Ohio's approval. The application for approval must include a waste minimization plan. The generator's implementation of the plan then becomes a condition of accepting the waste for disposal. Any generator that wishes to dispose of wastes in Ohio must, therefore, either develop and implement a waste minimization plan or demonstrate that it can achieve no practical reduction in its waste volume.

Other states may implement such mandatory hazardous waste reduction regulatory programs in the future. Federal law requires that, before a state can receive any Superfund money for the cleanup of hazardous waste sites, the state must be able to assure the EPA that adequate disposal capacity exists for all hazardous wastes generated within its borders.[95] Hazardous waste reduction programs are one mechanism states can employ to ensure that they will have sufficient capacity. Thus, this requirement may serve as an incentive for some states to adopt regulations mandating waste reduction. For example, New York is developing a program mandating that a generator that applies for a hazardous waste management permit must submit a Waste Reduction Impact Statement as part of the permit application. The statement will identify potentially applicable waste reduction techniques and contain a schedule for implementing a waste reduction program. Implementation of the program would then become a condition of the permit.[96]

State Educational and Technical Assistance Programs

Most states have not adopted regulatory programs but have instead encouraged hazardous waste reduction by providing assistance to the voluntary efforts of industry.[97] A large number of states have developed programs intended both to increase awareness of the benefits of waste reduction and to make available information about waste reduction techniques. These programs are generally run through the state's Office of Hazardous Waste Management. To increase awareness, these offices conduct seminars and workshops, send mailings, and otherwise try to "spread the word" about waste reduction. A number of states also have a Governor's Award that is given to companies that succeed in making substantial reductions in their hazardous waste generation.

These state agencies also run a variety of educational programs. Most serve as libraries or clearinghouses for technical information on waste reduction techniques. A number of states, including Connecticut, Kentucky, Maryland, New Jersey, and New York, publish newsletters or journals on waste reduction. States are also developing computerized databases of waste reduction information and computer-aided instruction programs on waste reduction. A few states, including California, Illinois, and New York, even fund research into waste reduction techniques.

Closely related are state technical assistance programs. These programs also are generally administered by the state Office of Solid or Hazardous Waste Management. In some states, however, these programs are either administered by a separate nonregulatory state agency or through a university. These programs are designed to assist individual companies to implement waste reduction techniques. Although in some states any generator is eligible for assistance, the program in many states is focused on either small businesses or small quantity generators.

Although some states offer only telephone or in-office assistance, most states with technical assistance programs provide on-site assistance. These states will send someone from the program to do an audit of the generator's facility, identifying areas where waste could be reduced and working with the generator on implementing a waste reduction plan. These programs have been able to achieve substantial cost savings for industry. Perhaps the largest such program is in North Carolina. Since it began in 1983, the program is estimated to have saved generators in the state more than twelve million dollars.[98] Participating facilities have, on average, reduced their annual volume of hazardous waste generation by more than 30 percent.[99]

Although the North Carolina program (and most others) responds to companies looking for assistance, the Massachusetts Department of Environmental Management has taken a more proactive approach.[100] The department studied waste generation in the state and determined that the electroplating and metal finishing industries were prime targets for a waste reduction program. The department then went to the generating facilities, investigated waste streams, and offered assistance in implementing waste reduction techniques. A number of the companies have implemented waste reduction programs and realized substantial savings.

Waste Exchanges

A number of states (including Alabama, Arizona, California, Indiana, and New York) have waste exchange programs. In addition, a number of regional waste exchanges are in existence. The purpose of a waste exchange is to match buyers and sellers of waste material with potential economic value. For example, a company interested in reclaiming mercury could use one of these exchanges to locate mercury-bearing wastes. The exchange receives listings of available waste and of buyers looking for recycled wastes. The information is then either put on a computerized database or published in the form of a listing. The goal of these exchanges is to increase the market for recyclable products and offer companies the opportunity to sell or give away waste for which they would otherwise have to incur hazardous waste disposal costs.

Economic Incentives

A number of states have adopted economic incentives to spur hazardous waste reduction efforts. Some states (including California, Connecticut, Illinois, Minnesota, North Carolina, and Oregon) have programs to give grants for waste reduction activities. In general, the grants are given for the development or demonstration of new waste reduction technologies.

Connecticut and New York both offer loans for purchase of pollution control equipment. These loans could be used for waste reduction activities.

Connecticut and Oregon give tax credits for the purchase of pollution control equipment, which also could be applied to waste reduction equipment. Oklahoma and North Carolina offer tax credits specifically for expenditures on waste reduction or recycling activities. Wisconsin offers a sales tax exemption on purchases of equipment for waste reduction.

A final economic incentive used by states is the fee often imposed on generators for disposal of hazardous waste. The existence of a fee is an incentive because the less waste that is generated, the less of a fee is paid. In addition, a number of states (for example, New Hampshire) structure their fees so that the amount of the fee depends on the destiny of the wastes: Wastes being landfilled are subject to the highest fees, wastes being treated are subject to lower fees, and wastes being recycled are exempt from fees.

In summary, state governments are heavily involved in the promotion of hazardous waste reduction activities. Although states generally do not yet require waste reduction activities, they are a potential source of information and assistance for a company wishing to implement a hazardous waste reduction program.

Recent Federal Legislative Initiatives

The current direction of federal policy on industrial waste reduction is best illustrated by the debate that surrounded House of Representatives 2800, the Hazardous Waste Reduction Act, in the 100th Congress and continuing discussion of that bill's reincarnation as House of Representatives 1457 in the 101st Congress. House of Representatives 2800 was introduced on June 25, 1987. The premise of this bill is that nonregulatory legislative measures are the most appropriate means to facilitate the reduction of industrial waste. This premise had been supported by work done by the National Academy of Sciences;[101] INFORM, a nonprofit environmental policy research organization;[102] the Congressional Office of Technology Assessment (OTA);[103] and the EPA.[104]

The provisions contained in House of Representatives 2800 would:

- Define waste reduction practices as those that prevent the generation, emission, or discharge of hazardous wastes at their sources
- Make matching grants to states for providing technical assistance to businesses seeking to exploit waste reduction opportunities
- Establish a national clearinghouse on waste reduction and recycling techniques that will serve as a center for waste reduction technology transfer
- Coordinate and improve the collection of information about waste reduction and recycling practices under RCRA
- Establish, maintain, and analyze a national database of information about waste reduction practices and opportunities for specific industries
- Report to Congress on progress toward achieving industrial waste reduction, on regulations that may inhibit waste reduction, and on priority pollutants and industries that could benefit from comprehensive waste reduction strategies

- Establish an Office of Waste Reduction to carry out the functions listed above, and to review for the Administrator the impact of any proposed environmental regulations on waste reduction[105]

The sponsors of this legislation (and the National Academy of Sciences, INFORM, EPA, and OTA reports that preceded it) concluded that significant opportunities exist for manufacturing industries to reduce chemical waste generation at little or no cost and that the nonregulatory federal initiatives proposed in the bill would facilitate such reduction.

The reporting provisions of House of Representatives 2800 in particular sparked controversy. These provisions, contained in Section 3 of the bill, would affect companies required to report releases of toxic chemicals pursuant to Section 313 of the Superfund Amendments and Reauthorization Act (SARA).[106] The Toxic Release Inventory (TRI) compiled from the first set of reports, which were submitted in 1987, was released by the EPA in June 1989.[107] According to this inventory, 22 billion pounds of 330 toxic chemicals were released to air, water, or land. The chemical product industry (SIC Code 28) accounted for more than half of all these releases (12 billion pounds).[108] Discharges to surface water (9.6 billion pounds) were about triple the size of the next largest category of releases.[109]

Only companies within SIC Codes 20 through 39 are required to report under Section 313 of SARA.[110] Within those manufacturing industries, only businesses making more than 25,000 pounds each year of a listed chemical, or using more than 10,000 pounds each year of a listed chemical, are required to report.[111]

House of Representatives 2800 would have required those businesses that were already required to report releases of a chemical under Section 313 to report, in addition, on waste reduction for each toxic chemical. This reporting would have included a calculation of the level of waste reduction achieved at a facility during the year in question and a comparison with the level of waste reduction achieved in the previous year. Estimates would also have been required of the level of reduction expected for the forthcoming year and the coming three years. A similar requirement would have been established for recycling.[112] Finally, House of Representatives 2800 would have required facilities to provide a production index for each chemical, so that waste reduction accomplishments could not be disguised or inflated by basic changes in production output. The details of the index to be calculated would be left to each business.

Varying opinions of House of Representatives 2800 were aired in testimony before the House Subcommittee on Transportation, Tourism, and Hazardous Waste on April 21, 1988. The American Petroleum Institute (API) and the Chemical Manufacturers Association (CMA) opposed the bill. Witnesses for both stated that the legislation was not necessary.[113] The API expressed concern that the bill would complicate and duplicate the reporting requirements developed under RCRA, stated that wastes should be tracked by volume rather than by chemical, expressed concern over indexing of production, and stated that treatment, such as incineration, should be included in the definition of waste reduction.[114] CMA had essentially similar concerns.[115]

The EPA did not support the bill. Instead, House of Representatives 2800 had the support of the nonprofit environmental community,[116] OTA,[117] an association of state agencies,[118] various elected and appointed governmental officials,[119] and Waste Management, Inc.[120] Dow Chemical Company also expressed general support for the principles of the bill.[121] The Natural Resources Defense Council (NRDC), while supporting the bill as a "very important first step,"[122] also proposed enactment of a mandatory source reduction and recycling "performance standard."[123]

With some modifications, House of Representatives 2800 was reported from the Subcommittee and the full Committee and passed by the House in the fall of 1988.[124] The Senate passed a similar but not identical bill. Despite repeated efforts to reconcile minor differences, the bill died at the end of the 100th Congress.

With some additional modifications, the legislation was reintroduced on March 15, 1989, by Representative Howard Wolpe[125] and Senator Frank Lautenberg.[126] The House bill, House of Representatives 1457, is titled the Waste Reduction Act; the Senate bill went a step further with the title, "Pollution Prevention Act of 1989." Differences between the bills are minor.

A hearing was held on House of Representatives 1457 on May 25, 1989, before the same House Subcommittee that conducted the hearing on House of Representatives 2800.[127] The tone of the members and witnesses was decidedly more conciliatory than in 1988. At the 1988 hearing, for example, Representative Wolpe was subjected to an unusually close examination by subcommittee members after he had presented his testimony on the bill.[128] This did not occur in 1989. Dow Chemical for the first time formally endorsed the bill at the 1989 hearing.[129] In a speech made earlier that month in London, the chairman of DuPont also endorsed the bill. Waste Management continued its support.[130] The CMA continued to raise reservations about the bill, but was generally more positive about its principles than in 1989.[131] Only the API continued to assert unequivocally that the legislation was unnecessary.[132]

The most dramatic shift in position was that of the EPA. EPA Administrator William K. Reilly testified that waste reduction legislation was needed.[133] Previously, the EPA had supported the principles of House of Representatives 2800 but took the position that its objectives could be accomplished through administrative action. In fact, before the May 1989 hearing, EPA had established a Pollution Prevention Office under the Assistant Administrator for Policy, Planning and Evaluation,[134] and inserted an optional request for waste reduction information in forms developed for reporting toxic chemical releases under Section 313 of SARA. The agency had also proposed a policy statement on pollution prevention.[135]

At the hearing, Reilly announced a fast track proposal to earmark 2 percent of the EPA's Fiscal Year 1991 extramural budget request for multimedia pollution prevention projects.[136] He also announced that the EPA was developing a multimedia pollution prevention strategy to make certain that pollution prevention is considered in base program activities.[137] Among the examples that he cited of actions under review was a plan for "monitoring and conducting toxicity reduction evaluations" in Clean Water Act permits.[138]

Administrator Reilly further indicated that the EPA would be submitting legislation that would "build upon" House of Representatives 1457.[139] The administration's proposal, he said, would "go beyond what is required in H. R. 1457—to encompass a wide range of generators of pollution (not just the manufacturing sector) and a wide range of pollutants released to all environmental media (not just toxic chemicals)."[140]

It appears likely that legislation incorporating the key features of House of Representatives 1457 will be enacted in the 101st, or perhaps the 102nd, Congress, possibly as part of broader RCRA reauthorization legislation. It is uncertain, however, whether the legislation will incorporate any new concepts introduced by the EPA or more draconian requirements such as those proposed by the NRDC. Toxic waste reduction provisions, for example, are gaining currency in state governments and, in the MSW field, are under serious consideration by Congress.[141]

MUNICIPAL SOLID WASTE

Federal Requirements

Federal activity in the area of municipal solid waste source reduction has, so far, been limited to the promotion of waste reduction efforts through broad policy statements, technical assistance, and procurement efforts. The EPA's Municipal Solid Waste Task Force, part of the Office of Solid Waste, for example, published an *Agenda for Action*, which underscores the need for increased efforts aimed at both source reduction and recycling of municipal solid waste. In addition, although the EPA's new Pollution Prevention Office deals primarily with hazardous waste reduction issues, information and assistance is also available from this office for municipal waste reduction.[142]

Financial assistance from the EPA is available to states and regional organizations under a grant program supporting efforts to develop or expand technical assistance programs that address the reduction of pollutants.[143] Although this grant program is not specifically directed toward municipal solid waste reduction, it could provide funding for an integrated waste program that includes MSW reduction.[144]

As with MSW source reduction, direct federal involvement in recycling under RCRA has been limited. One method the federal government has employed, however, involves the use of federal procurement; that is, the purchasing power of the federal government, to stimulate markets for recycled products.

Section 6002 of RCRA mandates that federal procuring agencies, when buying certain designated items, must purchase the items with the highest percentage of recovered materials.[145] This requirement applies to a federal agency only if: (1) the EPA has designated the item to which the requirement applies; and (2) the purchase price of the item exceeds $10,000 or the quantity of such items acquired in the preceding year was $10,000 or more.[146]

Aside from the responsibility to designate categories of items that are subject to the procurement requirements, the EPA must also issue guidelines designed to help the federal procuring agencies fulfill their obligations.[147] To date, the EPA has designated only a few items for this procurement preference and has published guidelines for each: (1) cement and concrete containing fly ash;[148] (2) paper and paper products containing recovered materials;[149] (3) lubricating oils containing re-refined oils;[150] (4) retread tires;[151] and (5) building insulation products containing recovered materials.[152] The EPA is currently studying whether procurement guidelines should be issued for additional categories. Items identified as possible candidates for federal procurement guidelines include automotive parts, remanufactured engines, and electronics.[153]

Procuring agencies that use federal funds and purchase $10,000 or more of the designated materials are required to develop programs, using the EPA guidelines, that maximize the procurement of items composed of the highest percentage of recovered material. The procuring agencies' programs must include a recovered material preference program; an agency promotion program; a program for requiring estimates, certification, and verification of recovered material content; and annual review and monitoring of the effectiveness of the procurement program.[154]

Federal procurement of products containing recovered materials is intended to demonstrate the technical and economic viability of these products. The EPA's procurement guidelines also provide guidance to state and local governments interested in procuring products containing recovered materials.[155]

In the most vigorous action to date in pursuit of its pollution prevention policy,[156] the EPA recently proposed a mandatory materials separation program for municipal solid waste incinerators, which the agency believes "would encourage source reduction and recycling of materials."[157] In a proposed Clean Air Act rule that regulates air emissions from municipal waste combustors (MWCs), the EPA has proposed a requirement that all MWCs would have to process municipal solid waste to achieve at least an overall 25-percent reduction by weight of MSW by separating for recovery some combination of the following: paper and paperboard, ferrous metals, nonferrous metals, glass, plastics, household batteries, or yard waste. MWCs would not be permitted to account for more than a 10-percent reduction through separation of yard waste. The proposed rule would also prohibit combustion of all household batteries and of lead-acid batteries weighing more than 5 kilograms. MWCs would be able to satisfy the 25-percent materials separation obligation through on-site manual or mechanical separation, through an off-site community source reduction or recycling program, or through a combination of these means.[158]

The EPA believes that this rule would reduce both the volume and toxicity of MWC air emissions and MWC ash.[159] The agency further believes that "the long-term viability of markets for recovered materials would be significantly enhanced by ensuring a stable supply of recovered materials."[160]

Substantial questions exist regarding the practicality of implementing this proposal. For example, where an MWC does not control the collection of the MSW it receives, it might have great difficulty in demonstrating that it had achieved the mandated reduction. One may also ask whether source reduction and recycling should be mandated as a means to develop market outlets for recovered materials or whether additional governmental initiatives aimed at market development should precede such action. Whatever the EPA's final answer to these questions, the MWC materials separation proposal marks a significant change in the direction of the EPA's policy on source reduction and recycling.

State Requirements

State Source Reduction Programs

Several state and local programs attempt to promote or mandate the reduction of municipal solid waste. These attempts have ranged from educational programs, which seek to instill an appreciation for the values of solid waste reduction within industry or the general public, to taxes and even bans on specific materials, product groups, or types of containers or packaging.

Florida is one of the states that has enacted the strongest of possible waste reduction measures, an outright ban on a particular type of material or product group. Florida legislation prohibits the use of plastic bags by retailers after January 1990, unless the plastic degrades within 120 days.[161] The use of polystyrene or plastic-coated paper on food packaging is also prohibited unless the material is certified as degradable by the Federal Food and Drug Administration.[162] In a similar action, the city of Portland, Oregon, has banned the use of polystyrene foam containers for prepared food by restaurants and retail food vendors starting January 1, 1990.[163] Nebraska has banned the sale of nonbiodegradable disposable diapers after 1992, and other states are considering imposing similar bans on disposable diapers.[164]

Outright bans have often faced strong industry opposition in the form of protracted litigation. In Suffolk County, New York, for example, an ordinance banning styrofoam food service packaging and other plastic wrapping has been tied up in court proceedings and has not been enforced.[165]

One type of outright ban that has proved successful and legally sustainable is a state or local ban on nonbiodegradable plastic six-pack rings. The success of these measures is largely due to the harm these

products inflict on wildlife and the ready availability of substitutes.[166] An example of a state that has eliminated production of this waste is Minnesota.[167] Similar actions include recent bans on beverage containers made of both plastic *and* steel or aluminum by several states, including Connecticut.[168]

States have also enacted partial bans, such as bans that apply only to government itself. Minnesota, for example, has prohibited the purchase of nonbiodegradable plastic trash bags by all public agencies.[169]

Another approach to waste reduction is to impose fees on producers or distributors of certain types of products. Florida, for example, has enacted an advance disposal fee of one cent for each container to be imposed on distributors of containers that are not recycled at a sustained rate of 50 percent as of October 1, 1992.[170] Florida has also imposed a product waste disposal fee on newsprint producers and publishers, which is offset by a credit for newsprint returned for recycling.[171]

Some municipalities have worked toward waste reduction through imposition of variable user charges for MSW disposal. Seattle, Washington, has such a program, in effect since 1981, where the customer decides upon a weekly level of service. Present costs are $13.55 for a single can to $31.55 for three cans a week.[172] Seattle recently added a mini-can service to promote further waste reduction and recycling.

Other measures to reduce the volume of MSW include a Rhode Island law requiring retailers that use plastic bags to also provide paper bags and post a sign to this effect.[173]

Legislative efforts aimed at source reduction of municipal waste have increased in recent years. The above listing is just a sample of the measures states have taken to combat their MSW management problems. Although the success of these programs is difficult to measure, increased consumer and political awareness is likely to foster further efforts. Moreover, the ongoing debate and heightened public involvement will probably encourage manufacturers to give more serious consideration to voluntary source reduction measures.

State Recycling Programs

State governments have also been increasingly active in efforts to encourage, and even require, the development and implementation of comprehensive recycling programs. A recent survey revealed that in 1989 alone, thirty-eight states and the District of Columbia enacted more than 120 recycling laws, and that twenty-seven states and the District now have comprehensive recycling laws in place.[174] Not surprisingly, most of this legislative action has been in the Northeast, where high population density, limited remaining landfill space, and high property values have combined to produce the most severe MSW disposal capacity problems.

State initiatives on the recycling front range from state-wide programs mandating source separation and curbside recycling, to technical assistance and grants for the development of pilot recycling programs, to educational and public awareness programs. States that have enacted mandatory source separation laws include Rhode Island, New Jersey, Connecticut, New York, Maine, Washington, and Pennsylvania. These mandatory programs vary in specific requirements and means of implementation.

New Jersey's mandatory recycling law is fully operational. Under New Jersey law, counties and municipalities must work with the state to achieve a minimum goal of recycling 25 percent of the solid waste stream.[175] Many New Jersey municipalities have adopted source separation ordinances and are providing for the collection of various recyclables, usually newspaper, glass, and aluminum cans from residences, and paper and corrugated cardboard from the commercial sector.[176] New Jersey also provides technical assistance to businesses and municipalities for the development of recycling strategies and operating procedures, and provides financial assistance in the form of grants, low-interest loans, tax incentives, and contracts for market development, program implementation, and education.[177]

Another example of a comprehensive mandatory MSW recycling program is that of Connecticut. Connecticut law mandates that each municipality or region must have in place, by January 1, 1991, a service that will ensure the recycling of cardboard, glass and metal food containers, leaves, newspaper, office paper, scrap metal, storage batteries, and waste oil.[178] The law requires municipal planning or municipal participation in a regional planning process. Plans are submitted and approved by the state environmental agency.[179] Many areas of Connecticut already have their recycling programs in place and operating.[180]

In July of 1988, Pennsylvania enacted a mandatory recycling bill that requires municipalities with a population of more than 10,000 to begin recycling by September 1990, and municipalities with a population of 5,000 to 10,000 and a population density of more than 300 persons per square mile to begin recycling by September 1991.[181] This law requires that communities recycle at least three of the following: glass, aluminum, other metal cans, office paper, newsprint, cardboard, or plastics.[182] According to a Pennsylvania official, 154 municipalities were already in compliance less than a year after the legislation passed and well before the first statutory deadline.[183] New York's Solid Waste Management Act of 1988 also requires mandatory source separation as of September 1992.[184]

Oregon has taken a slightly different approach. Although the state does not require that all citizens recycle, the Recycling Opportunity Act of 1983, which took effect in 1986, mandates that citizens be given the *opportunity* to participate in curbside recycling.[185] This law applies to municipalities of 4,000 or more.[186] The state environmental agency determines which materials are recyclable on the basis of economic criteria for each waste management region or *waste-shed*. Waste-sheds are required to submit an original plan and annual reports describing local implementation of curbside recycling and to conduct a mandatory education and promotion program.[187]

Other states have also passed recycling measures that stop short of requiring mandatory recycling by citizens, but still direct that local governments include recycling measures in their planning processes and eventually implement them. Virginia, for example, has recently enacted legislation that requires local development of solid waste management plans that must include a component describing how local governments will meet recycling goals of 10 percent by 1991, 15 percent by 1993, and 25 percent by 1995.[188]

Recent legislation in Florida requires countywide recycling programs that increase the amount of waste being recycled from a current level of 10 percent to 30 percent by 1994.[189] Illinois enacted a recycling law in August 1988, which requires Chicago and counties with more than 100,000 residents to develop waste management plans that emphasize recycling and landfill alternatives.[190] The plans must be designed so that 25 percent of municipal waste is recycled by the fifth year of the plan.[191]

In December 1988, Michigan passed a series of recycling laws, which include a requirement for updated five-year county-wide solid waste plans that must provide for recycling or establish that recycling is not necessary or feasible.[192] Given Michigan's position that at least some recycling is feasible in every county, this law can be seen as requiring the inclusion of recycling at least at the planning level.[193] It is hoped that the law will go a long way toward reaching the statewide goal of 50-percent recycling of the solid waste stream by 2005.[194]

Delaware has opted not to pursue source separation or curbside recycling, but instead has a large-scale materials processing program operated by the Delaware Solid Waste Authority.[195] This front-end system in Newcastle County services 70 percent of all households in the state. The processing facility recovers glass, aluminum, and ferrous metals; produces compost from organic material and sewage sludge; and produces RDF (refuse-derived fuel), which is burned to produce steam and electric power.[196]

Some states are linking the permitting of new solid waste facilities, such as mass-to-energy incinerators, with local development of recycling capacity. Massachusetts, for example, has imposed a moratorium on the permitting of new waste-to-energy incinerators.[197] After a review of state disposal capacity, permitting will proceed only where proposals meet criteria that ensure a certain level of recycling.[198] New York's solid waste regulations require that applicants for landfill or incinerator permits, including renewal permits, submit and obtain approval of a comprehensive recycling analysis, unless such a plan is already in effect. The recycling analysis must include identification of the quality and types of recyclables, including strategies for source reduction; an evaluation of existing efforts to recover recyclables; the identification of markets for recovered recyclables; and a plan for the implementation of a recycling program.[199]

Some states, although not mandating recycling or even requiring the inclusion of recycling in the solid waste planning process, nevertheless are actively moving down the recycling path by making funds available for the development of recycling programs. The state of Washington, for example, has announced that it will provide more than $16 million to local governments for planning, feasibility studies, design, construction, and administration of recycling programs.[200] Other states that have funding programs include Illinois, where the state environmental agency has the authority to provide low- or zero-interest loans for recycling efforts by businesses, not-for-profit groups, and governmental organizations. Michigan has established a plastics recycling development fund and consortium to provide grants and loans for the funding of new recycling efforts aimed specifically at plastics.[201]

Other measures to encourage recycling include procurement programs that aid the development of markets for recyclables. Illinois[202] and Florida[203] are among those that have adopted measures requiring such procurement programs. To aid in the recycling of plastics, some states are also mandating the coding of plastic containers according to resin type. Examples of states that have enacted such measures include Michigan[204] and Illinois.[205]

Another approach used to reduce solid waste is through beverage container laws. During the 1970s, several states enacted mandatory beverage container deposit laws. Although these bottle bills were primarily designed as a way to control litter, they could also be promoted as a way to reduce waste. These laws require the consumer to pay a deposit on beverage containers upon purchase. Empty bottles are returned by the purchaser to retailers, where the deposit is redeemed. Mandatory beverage container deposit laws are in place in at least eight states, where they typically achieve return rates of 85 to 90 percent.[206] Other states have established voluntary Beverage Industry Recycling Programs (BIRPs). These programs simply urge consumers to deliver cans, bottles, and plastic containers to buy-back centers, where they can receive payment and the materials can be prepared for market.[207]

In summary, many states have adopted far-reaching measures to reduce the amount of MSW requiring disposal. As discussed in the next section, this trend is likely to gain impetus from initiatives now under consideration by Congress.

Recent Federal Legislative Initiatives

Several bills promoting source reduction and recycling of MSW were introduced during the 100th Congress. Foremost among these in depth and potential momentum was Senate 2773, a comprehensive bill to reauthorize RCRA, introduced by Senator Max Baucus and others on September 9, 1988.[208] Senate 2773 and the other bills died at the end of that Congress, but the Baucus bill, with some revisions, was reintroduced in the 101st Congress.[209] This bill, together with waste reduction legislation introduced on the same day by Senator John Chafee,[210] are the principal Senate legislative proposals on waste reduction in the 101st Congress.

A companion overall RCRA reauthorization bill was introduced in the House in the first session of the 101st Congress,[211] by Representative Tom Luken, Chairman of the House Subcommittee on Transportation and Hazardous Materials, who set an ambitious schedule of hearings for 1990. Several House bills specifically addressing waste reduction alone were also introduced. The Recycling Promotion Act, introduced by Representative Ron Wyden, is particularly significant.[212]

The specific provisions of the Baucus, Chafee, Luken, and Wyden bills are not likely to be enacted without substantial revision. Nevertheless, these four major proposals each incorporate several concepts that are being discussed extensively by interested parties throughout the country, and these concepts are likely to be incorporated into law in some form in the foreseeable future. These concepts are summarized below.

Waste Reduction Goals
The Baucus and Chafee bills would establish ambitious national

waste reduction goals to be implemented by state governments.[213] The Baucus and Chafee bills would mandate a national goal of 25-percent municipal solid waste recycling within four years and 50-percent recycling of MSW within ten years. The Baucus bill would limit the 50-percent goal to circumstances where recycling constitutes "least cost disposal." A goal of 10-percent municipal waste reduction within four years would also be required, although the Baucus bill provides an exemption to this requirement if the state in question demonstrates that the reduction "is not practicable."[214]

To reach the 25-percent recycling goal would require recycling of an additional 22 million tons of MSW over the amount of waste currently recycled. This calculation takes into account estimates that 17 million tons (about 11 percent) of MSW are now recycled. Recycling an additional 14 percent appears to be achievable in many locations if yard waste is collected for composting and programs are established for curbside collection and subsequent processing of paper, glass and plastic bottles, and metal cans.

The percentage recycling and waste reduction goals of the Baucus and Chafee bills would be enforced by a carrot, rather than a stick. Under both bills, eligibility for EPA solid waste grants is limited to those states that have adopted solid waste management plans and that are making progress on waste reduction and recycling.[215]

The Luken bill does not purport to establish a goal or to prohibit grants to states or communities that have *not* made progress on waste reduction. Instead, it offers a recycling incentive by authorizing the EPA to make bonus grants, which can be used for any municipal purpose, to any municipality or other agency that demonstrates a recycling rate greater than 25 percent.[216]

Fee on Nonrecycled Packaging Materials
Appropriations for solid waste management activities have been very limited, and those concerned with the next iteration of solid waste legislation are interested in mechanisms that would earmark funding for this purpose. The Chafee and Luken bills envision a system of fees attaching to use of virgin materials.[217] The fee proposed in special companion legislation to the Luken bill would attach to

virgin material used in a broad range of products.[218] This fee would be $7.50 a ton of virgin materials.[219] Those interested in this legislation estimate that these fees would raise several hundred million dollars annually, which would be directed to state government waste management programs.

Senator Chafee had drafted but did not introduce companion legislation similar to Representative Luken's. Senate 1112 nevertheless retains reference to a Source Reduction and Recycling Trust Fund to be funded under separate legislation.[220] Among the uses of this trust fund authorized by the Chafee bill would be tonnage grants, allocating 35 percent of the funds raised to recycling grants, which would be distributed to states on the basis of the tonnage of materials recycled within each state.[221]

Limits on Manufacture and Disposal

The Baucus, Chafee, Luken, and Wyden bills all contain provisions that would authorize the EPA to limit the manufacture and distribution of products if their contribution to MSW toxicity or volume is found to be excessive. Heavy metals are the most obvious target of the toxicity reduction provisions, while the provisions addressing volume reduction focus heavily on excess packaging.

Toxicity

The Baucus bill includes a section on hazardous constituents in products that would require the EPA to publish a list of no fewer than ten products, augmented by ten more products each year, and to identify in those products hazardous substances presenting risks to human health and the environment when disposed or incinerated.[222] The administrator of the EPA would be authorized to issue regulations governing disposal or incineration of listed products as may be necessary to protect human health and the environment from these risks. If the administrator determined that these regulations would not adequately protect against the risks posed by a product, he would be required to set a schedule, not exceeding two years, for review of the product under Section 6 of the Toxic Substances Control Act.[223] That statute provides the EPA with authority to ban the manufacture, processing, or distribution of products if they "present an unreasonable risk of injury to health or the environment."

The Luken bill calls on the EPA to identify, in each of the five years following enactment, five "of the most toxic and common constituents of municipal waste."[224] Within two years of identifying each of these twenty-five constituents, the EPA is to consider imposing (1) a ban on the use of each constituent in production; (2) a ban on the land disposal of products containing each constituent; (3) special management standards; or (4) a mandated use of substitute constituents. The EPA could not impose, however, any ban on the use of a constituent in production or any requirement to use a substitute constituent in production except under the authority of other statutes and not under RCRA.

The Chafee bill also addresses hazardous constituents in products, but is more stringent than the Baucus or Luken proposals. Section 108(b) would similarly direct the EPA to issue regulations necessary to ensure that the disposal or incineration of products will not threaten human health and the environment. Under the Chafee legislation, however, the regulations would include prohibitions or limitations on the manufacture, processing, or distribution of such products and on allowable concentrations of substances in such products. Thus, while the Baucus and Luken bills would leave final disposition of any product ban or limitation to the EPA's existing statutory authority, the Chafee bill would create new authority for the EPA to regulate production processes under the RCRA standard of protection of human health and the environment. Additional authority is provided elsewhere in the Chafee bill for the EPA to issue national packaging standards that among other things, "eliminate, to the greatest extent practicable, the use of toxic materials in packaging."[225] Finally, the Chafee bill would impose an outright ban on the use of cadmium as a pigment and on all other "nonessential" uses of that metal.[226]

The Wyden bill in the House would authorize the EPA to prohibit commonly used constituents of consumer packaging if it determines that "the constituent as used makes resource recovery or disposal of the packaging more difficult or dangerous because the constituent has a characteristic of hazardous waste."[227] If the EPA administrator makes such a determination and also determines that the constituent is not necessary to "safe and healthful packaging," the EPA could issue regulations prohibiting the use of the constituent in packaging.

Volume

The Baucus, Luken, and Wyden bills do not provide authority for the EPA to mandate volume reductions, but such authority is contained in provisions of the Chafee bill. That bill would direct the EPA administrator to issue regulations establishing national packaging standards to, among other things, "minimize the quantity of packaging material in the waste stream." The regulations would be due within two years of enactment of the provision and would become effective within five years of issuance.[228] Packaging manufacturers would be required to comply with the standards.

Labeling

The Chafee, Baucus, and Wyden bills would establish labeling requirements for recycled goods in general and for plastics in particular. The Luken bill takes a narrower focus. The Chafee bill would require the administrator to develop a national recycling seal or symbol and to identify requirements that must be met before an item could bear this symbol, which would characterize it as recyclable or as containing recycled material.[229] A similar provision in the Baucus bill would be implemented by the packaging committee discussed below.[230] The Chafee bill and the Luken bill would each require that plastic packaging and products bear a standardized label identifying the plastic resin used to produce the product.[231] In addition, the Chafee bill would require labeling of any product or article if necessary to alert consumers to the presence of hazardous constituents that may present a risk to human health or the environment.[232]

The Wyden bill contains labeling provisions similar to those described above. In addition, the bill would require packaging to be labeled as "nonrecyclable" if the administrator finds that waste generated from it is not readily recoverable because of the packaging's "design, constituents, or other related reasons."[233]

Packaging Review Committee

The Baucus, Chafee, and Luken bills would establish an advisory commission or packaging institute or board to evaluate the effect of standards and practices in packaging on waste reduction and recycling. The Chafee proposal would set up a Products and Packaging Advisory Board,[234] while Representative Luken's bill calls for a Presidential Commission on Waste Reduction.[235] These bodies would report to the EPA administrator on minimizing the quantity of packaging and on other matters, including the labeling requirements discussed above. The Baucus bill would establish a considerably more elaborate National Packaging Institute.[236] The Institute would have essentially the same mission as the advisory bodies discussed above, but would operate as a quasi-independent agency in advancing that mission. For example, the Institute would be directed to establish (not just recommend to the EPA) national packaging standards and to decide whether a seal or symbol developed by the Institute could be used on particular package types.

Waste Separation Requirements

Of the major legislative proposals discussed here, only the Chafee bill contains a variation on the theme of mandatory source separation. The Chafee bill sets forth various requirements for state plans, which must be implemented unless the administrator believes that a requirement is not practicable.[237] Among these, states must require that "recyclable solid waste from commercial establishments and office buildings be separated prior to deposition in municipally owned or operated" facilities. A surcharge on tipping fees would be imposed if no such separation were made.

Management of Batteries

All four bills contain stringent, but different provisions concerning batteries in the MSW stream. Lead-acid batteries are of particular concern. In the 1960s, the lead in more than 95 percent of such batteries, which are used in cars, trucks, and buses, was returned to secondary lead smelters where it was processed for reuse into new batteries. Today, only about 80 percent of these batteries are recycled, leaving an estimated 138,043 tons of lead—about 65 percent of MSW lead discards in 1986—outside of the recycling chain.[238]

The Baucus bill would flatly prohibit the land disposal or incineration of lead-acid and mercury batteries, and would direct the EPA to issue whatever regulations are necessary to protect public health and the environment from the hazards associated with recycling these batteries. The Baucus bill would also authorize issuance of special regulations for lead-acid and mercury battery recycling and would limit Superfund liability for releases of hazardous substances associated with management of these batteries.[239]

The Chafee bill would also ban the landfilling or incineration of lead-acid batteries.[240] The bill would also prohibit disposal of such batteries except by delivery to an automotive battery retailer or wholesaler, a permitted secondary lead smelter, a state-approved collection facility, or the EPA. Any person selling lead-acid batteries would be required to accept as many old batteries from customers as the number of new batteries that are purchased. In addition, the Chafee bill would establish labeling requirements for lead-acid batteries.

The Wyden bill would establish a general take-back system, which would be expressly applicable to lead-acid batteries, for recycling of products: (1) that may constitute a hazard to human health or the environment if improperly disposed or improperly recovered, and (2) that are likely to be improperly disposed or improperly recovered if special requirements are not imposed. The take-back system would require the "retailer, distributor, importer, and manufacturer of the product" to "accept, at point of transfer, reasonable quantities of the used product from their customers without charge."[241] The administrator would be required to issue regulations for the operation of such a system for lead-acid batteries. Within three years thereafter, he would be required to determine if similar take-back systems are warranted for tires, dry-cell batteries, used oil, large household appliances, automobiles with air bags, and unused pesticides.

The Luken bill combines features of the Chafee and Wyden bills. This bill makes it a crime to dispose of a used lead-acid battery except by recycling as prescribed in the bill.[242] As under the Chafee bill, the permissible means of disposing of used lead-acid batteries would be limited to delivery to a battery retailer or wholesaler, a secondary lead smelter, or a collection or recycling facility permitted by a state or the EPA. Retailers and wholesalers would be required to take back used batteries. Finally, the EPA would be required to study and report to Congress on the health and environmental effects of used household dry-cell batteries in MSW and the potential recyclability of these products.

Federal Agency Management

Procurement

Virtually everyone involved in legislative efforts to promote recycling has recognized the priority of legislation to ensure that sound markets exist for items diverted from the waste stream. In the past, newspapers collected by Boy Scouts have sometimes been landfilled for lack of buyers. In 1988, the market price was high for newspaper and other key diverted wastes. In 1989, however, the price offered for old newspapers dropped sharply. Those who collect newspapers in the Northeast now must often pay to have the newspapers taken off their hands. High avoided disposal costs still make recycling of newspapers attractive economically in many areas. Nevertheless, the potential exists for an even worse market glut of newspaper and other items.

Federal agency procurement of recycled items is one tool for countering this problem.[243] Federal procurement can serve both as a substantial source of market demand in our economy and as a demonstration project for other buyers. Recognizing this potentially strong force, all four legislative proposals contain provisions on federal agency procurement of recycled items.

A major problem with existing EPA procurement guidelines for paper is the definition of what is an unreasonably higher price for paper with recycled content; i.e., the price at which a federal agency is not required under RCRA to buy recycled paper. The EPA has defined this price as anything greater than the price of paper made from virgin material—even one penny.[244] All four bills would amend

RCRA to define an unreasonably higher price for paper with recycled content as a price that exceeds alternatives by more than 10 percent.[245] These bills would require reissuance of the EPA paper procurement guidelines using this revised definition of what is an unreasonably higher price.

The Baucus bill contains provisions targeted at the postconsumer commercial and residential discards that are increasingly collected by curbside collection programs.[246] The bill would mandate the issuance of guidelines by set dates for the procurement of compost from yard waste and sewage sludge, plastic from discarded bottles, aluminum and steel from discarded cans, glass from discarded containers, used tire fragments in road cover, and recycled lead in batteries. The Baucus bill also contains hammers that would fall if the EPA misses the deadlines for issuing these guidelines. Federal agencies would be enjoined directly by the statute from authorizing, funding, or carrying out any action unless any of the designated products used in the action contain a specified minimum recycled content.

The Chafee bill similarly mandates issuance of procurement guidelines for various items. The items and deadlines differ from those of the Baucus bill, however, and no hammers are established.[247] The Wyden and Luken bills mandate issuance of procurement guidelines for at least three unspecified new items within two years after the bill is enacted. These bills also set forth detailed requirements for affirmative federal agency procurement programs regarding items covered by the guidelines.[248]

In addition to item-specific procurement requirements, both the Baucus and Chafee bills would require the use of recycled material in any federal contract for $1 million or more.[249]

Agency Waste Reduction Programs and Petitions

The Wyden bill would establish a broad duty on the part of all federal agencies to review existing and proposed regulations and to revise such regulations for the promotion of resource recovery.[250] Revisions for this purpose would be required unless they are not authorized by law, the cost of revision exceeds the value of any possible environmental benefit, or the revision would pose a significant threat to human health or the environment. A petition process is set forth in the bill pursuant to which any person could seek revisions in regulations or internal agency procurement specifications to promote resource conservation.

The Baucus, Chafee, and Luken bills also contain sections authorizing any person to petition federal agencies to take actions that would advance waste reduction and recycling.[251] *Action* is not defined but presumably could range from a specific procurement decision to an agency regulation of general effect. Under all three bills, the petitioner would be required to show that the effect of the action would bring about a 10-percent increase in recycled content or a 10-percent reduction in waste volume or toxicity; that the action is consistent with federal statutory requirements; and that the action would not cost the federal government money. If these three conditions are met, the federal agency would be obligated to take the action requested. The Baucus, Wyden, and Luken bills would require the designation of federal agency waste reduction or solid waste control officers who would be accountable for overseeing the petition process, for preparing agency waste reduction plans, and for reporting to the EPA on implementation of those plans.

The Wyden and Luken bills also contain a sweetener to encourage federal agencies to recycle. Executive agencies and units of the legislative branch are authorized to retain any funds they receive from the sale of recyclable material or from energy recovery from solid waste.[252]

Federal Assistance

Promotion of Commerce

The Wyden, Chafee, and Luken bills require the secretary of commerce to take steps that would promote markets for recovered materials. The Wyden and Luken bills direct the secretary of commerce, at least every two years, to survey the uses of recovered industrial and consumer materials and to identify and quantify markets.[253] These bills would also establish an interagency working group to undertake various tasks to promote increased use of recovered materials. The Chafee bill would require the secretary of commerce to submit biennial reports to Congress on an overall implementation of the bill.[254] Specific recommendations would be required on the need to continue or revise a program of low-interest loans for businesses that would be funded by federal grants.[255] The secretary would also be required to report on national and international markets for recyclable materials. In addition to these reporting requirements, the secretary of commerce would be directed to help exporters of recyclable items find good foreign markets and secure favorable financial terms. The Luken bill similarly would instruct the commerce secretary and the United States trade representative to identify foreign markets and assist exporters of recovered and recycled materials in selling to such markets.[256]

Financial and Technical Assistance

The Baucus, Chafee, and Luken bills would provide financial and technical assistance for source reduction and recycling. The Baucus bill authorizes grants to states from the funds discussed previously to assist "in developing and implementing a program to promote the use of waste reduction and recycling techniques by businesses, local governments, or regional waste management authorities."[257] The federal share of program costs would be limited to 50 percent. The Chafee bill would authorize similar grants "for the purpose of encouraging source reduction and recycling."[258] Thirty-five percent of funds dispersed would be for "recycling grants," whose size would be calculated on the basis of the total number of tons of material recycled from MSW on an annual basis within a state. Other grants would be available to states, local government, businesses, and other institutions. The Luken bill would authorize technical and financial assistance to states for waste reduction and recycling programs and, moreover, EPA loans to private persons for the capital costs of converting to greater use of recycled materials or for construction and operation of recycling facilities.[259]

Waste Reduction Office and Clearinghouse

The Baucus and Chafee bills would establish waste reduction[260] or

waste minimization[261] offices at the EPA and these bills, as well as the Luken bill, would set up clearinghouses within the agency to further waste reduction.[262] The Office of Waste Reduction established by the Chafee bill would be responsible for staffing the new EPA responsibilities authorized by the bill, which principally concern MSW. The Office of Waste Minimization established by the Baucus bill would similarly have lead staff responsibilities for newly created source reduction and recycling duties, and would cover industrial wastes as well as MSW. The clearinghouse established by all three bills would be structured to facilitate the exchange of information on source reduction and recycling of all kinds of solid waste. The Chafee bill would also mandate a focused educational effort on source reduction and recycling to be implemented through school curricula, public service announcements, new media campaigns, and information distributed at retail establishments.

Report to Congress

The Baucus bill would require the EPA to submit biennial reports to Congress on actions taken to implement the bill's waste reduction and recycling goals. Other reports would be required on industrial waste reduction.[263]

* * *

The likelihood of these federal policies being enacted is uncertain, as is the timing of any such action. We expect that many of the concepts discussed in this chapter will be incorporated into federal law, however. Moreover, some of those that are not enacted by Congress will likely be passed by state legislatures, if they have not already been made matters of state law. The policy and practice of waste reduction are a turn in the direction of society whose time has come.

ENDNOTES

1. 42 U.S.C. §§ 6901 *et seq.* The Solid Waste Disposal Act underwent major revision in 1976, with the enactment of the Resource Conservation and Recovery Act. This statute, among other things, established a comprehensive federal scheme for management of hazardous wastes, called for significant upgrading of state regulation of solid waste, and identified waste reuse and recovery as desirable national goals.

2. Section 1004(27) of RCRA defines the term *solid waste* to mean "any garbage, refuse, sludge, from a waste treatment plant, water supply treatment plant, or air pollution control facility and other discarded material, including solid, liquid, semisolid, or contained gaseous material resulting from industrial, commercial, mining, and agricultural operations, and from community activities, but does not include solid or dissolved material in domestic sewage, or solid or dissolved materials in irrigation return flows or industrial discharges which are point sources subject to permits under section 402 of the Federal Water Pollution Control Act, as amended (86 Stat. 880), or source, special nuclear, or byproduct material as defined by the Atomic Energy Act of 1954, as amended (68 Stat. 923)."

3. *See* Toxics Release Inventory, U.S. EPA, EPA 560/4-89-005 (June 1989) (hereinafter cited as "Toxics Release Inventory") (compiled under section 313 of the Superfund Amendments and Reauthorization Act, 42 U.S.C. § 11023). These data are summarized in *Danger Downwind: A Report on the Release of Billions of Pounds of Toxic Air Pollutants,* National Wildlife Federation (March 22, 1989). For further discussion of this inventory, *see* p. 73.

4. Press Briefing, U.S. EPA, *Municipal Compliance with the Clean Water Act* (July 27, 1988).

5. Office of Technology Assessment, U.S. Congress, OTA-334, *Wastes in Marine Environments* at 66, 67 (April 1987).

6. *Report to Congress: Solid Waste Disposal in the United States,* U.S. EPA, EPA/530-SW-8-011, Vol. I, at 11 (October 1988) (hereinafter cited as "EPA Solid Waste Report").

7. EPA Solid Waste Report, *supra* note 6, Vol. II, 3-31 (one billion gallons per day is approximately 1.4 billion metric tons per year).

8. Congressional Budget Office, *Hazardous Waste Management: Recent Changes and Policy Alternatives* at 18, Table 2 (May 1985) (hereinafter cited as "CBO Hazardous Waste Report") (estimating 1983 national hazardous waste generation to be 265.6 million tons); *The Hazardous Waste System* U.S. EPA, EPA/870-HW-8-002, ES-2 (June 1987) (estimating 1986 national hazardous waste generation to be 275 million tons).

9. Assuming 244 million people live in the United States, solid waste generation, not including agricultural wastes, would be 11.67 tons/year divided by 244 million people = 47.8 tons/person/year. If agricultural wastes are included and assumed to be generated at the upper limit of 1.4 billion tons/year estimated by EPA, per capital generation

would be 53.7 tons/person/year. Hence, 50 tons/person/year is a good, rounded approximation of per capita solid waste generation in the United States.

10.   The underlying statistics for these numbers and for trash are presented in Chapter II.

11.   For an extensive discussion of the environmental and economic incentives for waste reduction, *see*, e.g., Office of Technology Assessment, U.S. Congress, OTA-O-424, *Facing America's Trash: What's Next for Municipal Solid Waste* (October 1989) (hereinafter cited as "Facing America's Trash"); Office of Technology Assessment, U.S. Congress, OTA-317, *Serious Reduction of Hazardous Waste for Pollution and Industrial Efficiency* (1986) (hereinafter cited as "Serious Reduction"); Office of Technology Assessment, U.S. Congress, OTA-347, *From Pollution to Prevention: A Progress Report on Waste Reduction* (1987) (hereinafter cited as "Pollution to Prevention").

12.   Those embracing the concept include the Environmental Protection Agency (e.g., *The Solid Waste Dilemma: An Agenda for Action*, EPA/530-SW-89-019, at 17 (February 1989) (hereinafter cited as "Solid Waste Dilemma"); Chemical Manufacturers Association (e.g., *Hazardous Waste Reduction Act, Hearings on H.R. 2800 Before the Subcomm. on Transportation, Tourism, and Hazardous Materials of the House Comm. on Energy and Commerce*, 100th Cong., 2d Sess. 261 (1988) (hereinafter cited as "1988 Hearings") (statement of Mort Mullins)); Participants in the Keystone National Hazardous Waste Management Strategies Project, including individuals employed by Dow Chemical, the Sierra Club, U.S. PIRG, NRDC, Waste Management, Inc., and other organizations (The Keystone Center, *Source Reduction Report* (September 15, 1988)). On the other hand, the Office of Technology Assessment, in its recent report on the subject, embraces a simpler hierarchy, consisting only of waste "prevention" (i.e., source reduction) followed by waste "management" (i.e., everything else). Facing America's Trash, *supra* note 11, at 6-7.

13.   E.g., Ocean dumping; end-of-pipe discharges into air or water.

14.   E.g., Lead-acid batteries that are not recycled or sent to a waste management facility.

15.   Charles E. Roos, "Is Lead a Problem?" *Waste Age Magazine* (February 1988), cited in Solid Waste Dilemma, *supra* note 12, at A.F-11.

16.   EPA Solid Waste Report, *supra* note 6, Vol. II, 4-13, Table 4-10.

17.   EPA Solid Waste Report, *supra* note 6, Vol. I at 10, 13.

18.   *Characterization of Municipal Solid Waste in the United States : 1990 Update*, U.S. EPA, EPA/530-SW-90-042A (June 1990) (hereinafter cited as Characterization of MSW).

19.   EPA Solid Waste Report, *supra* note 6, Vol. I at 13.

20.   *See id.*, Vol. II, Table B-1.

21.   *Id.* at 3-22.

22.   *Id.* at 4-103.

23.   *Id.*, Vol. I, at 13.

24.   *Id.*, Vol. II, at 4-103.

25.   *Id.* at 3-32.

26.   *Id.* at 3-33.

27.   *Id.* at 4-103.

28.   *Id.* at 3-32.

29.   *Id.*

30.   *Id.*

31.   *Id.*

32.   CBO Hazardous Waste Report, *supra* note 8, at 17.

33.   *Id.* at 26.

34.   A very small amount is put to use as "lean water" to provide a heat sink in incinerators. A heat sink lowers the temperature in the incinerator kiln to a level that will not stress the kiln construction material.

35.   *See* 55 Fed. Reg. 11798 (March 29, 1990) (Toxicity Characteristic Leaching Procedure).

36.   EPA Solid Waste Report, *supra* note 6, Vol. II, at 3-29.

37.   *Id.* at 3-30.

38.   *Id.* at 3-15.

39.   *Id.* at 3-16.

40.   *Id.*

41.   *Id.* at 4-103.

42.   *Id.* at 3-31.

43.   *Id.* at 3-30.

44.   *Id.* at 4-103.

45.   Based on statistics in Characterization of MSW, *supra* note 18.

46.   *Report to Congress: Minimization of Hazardous Waste,* U.S. EPA, EPA/530-SW-86-33, at 45 (1988) (hereinafter cited as EPA Waste Minimization Report); Serious Reduction, *supra* note 11, at 16, 25; Pollution to Prevention, *supra* note 11, at 17-18.

47.   M.D. Konigsberger, *Pollution Prevention Pays,* presented in Nashville, TN (March 1986); *see also* 1988 Hearings, *supra* note 12, at 26 (statement of Rep. Claudine Schneider).

48.   Charles Rooney, "Managing Waste," *America Paint & Coatings Journal* (January 2, 1989).

49.   *See* 54 Fed. Reg. 3845, 3847 (Jan. 26, 1989); *Hazardous Waste Reduction Act: Hearings on H.R. 1457 before the Subcomm. on Transportation and Hazardous Materials of the House Comm. on Energy and Commerce,* 101st Cong., 1st Sess. at 31 (1989) (hereinafter cited as 1989 Hearings) (statement of William K. Reilly, Administrator of EPA).

50.   RCRA was amended in 1984 to ban the land disposal of hazardous wastes unless they have first been treated to reduce the toxicity or mobility of their hazardous constituents. 42 U.S.C. § 6924(d), (e), & (g). Congress mandated that, for each hazardous waste regulated under RCRA, EPA establish treatment standards that must be met before the waste, or its treated residue, can be land disposed. *Id.* § 6924(m). In implementing this statute, EPA has chosen to require that wastes be treated to a level achievable by the Best Demonstrated Available Technology (BDAT) for that waste.
        The ban on land disposal is being implemented in several steps. The first group of restrictions went into effect in November 1986 and all the restrictions will be in effect by May 1990.

51.   Taken from an article by Ryan Delcambre, "Waste Reduction: Program, Practice, and Product in Chemical Manufacturing," *Environmental Progress,* Vol. 7, No. 3, at 175-79 (August 1988).

52.   Characterization of MSW, *supra* note 18.

53.   *Id.*

54.   Reasonable people can differ on whether a diaper is a good or a container.

55.   Solid Waste Dilemma, *supra* note 12, at 2.A-6.

56.   Solid Waste Dilemma, *supra* note 12, at 1-12; *see* pp. 86-87, 89, 90.

57.   *Id.*

58.   *See* William L. Rathje, "Rubbish," *The Atlantic,* Vol. 264, No. 6, at 99, 102-03 (December 1989).

59. Distributed to participants at a Conservation Foundation dialogue on June 14, 1988, on the subject of "design for disposal" policy.

60. The most popular of these pigments are Cobalt Aluminate Blue, Chromium Antimony Titanate, and Nickel Antimony Titanate.

61. The most widely used lead chromate pigments are Chrome Yellow, Chrome Orange, and Molybdate Orange.

62. Cadmium pigments include Cadmium Sulfide Yellow, Cadmium Sulfoselenide Red, and Mercury Red.

63. W.D. Rinehart, American Newspaper Publishers Association, *Newspaper Ink* (March 16, 1988).

64. *See* pp. 87-89.

65. The Sonoma County program is conducted by Empire Waste Management, a division of Waste Management, Inc. Waste Management currently conducts recyclable materials collection programs in 142 communities.

66. 1986 Self-reported program participation data. Sources: Resource Conservation Consultants; *Waste Age Magazine*; Ferrand-Scheinberg Associates.

67. Data for studies performed in five communities. Source: Resource Conservation Consultants.

68. Characterization of MSW, *supra* note 18.

69. National Association of Recycling Industries, Rubber Recycling Division.

70. *See* 54 Fed. Reg. 3845, 3847 (Jan. 26, 1989); 1989 Hearings, *supra* note 49, at 31 (statement of William K. Reilly, Administrator of EPA).

71. *See* note 1 *supra*.

72. *See* 51 Fed. Reg. 10146, 10168 (Mar. 24, 1986). Note, however, that some states have not adopted this EPA interpretation of the RCRA rules and will require a permit except for certain limited forms of treatment expressly exempted in the regulations (e.g., neutralization or addition of adsorbent material).

73. Facilities in existence when the RCRA hazardous waste regulations became effective that have applied for, but not yet received, a RCRA permit have "interim status" and may operate as if they had a permit.

74. 42 U.S.C. § 6922(b); *see* 40 C.F.R. Part 262, Appendix.

75. 42 U.S.C. § 6922(a)(6)(7); 40 C.F.R. § 262.41(a)(6)(7).

76.   42 U.S.C. § 6925(b); 40 C.F.R. § 264.73(b)(9).

77.   A sludge is defined under RCRA as a waste generated by a waste treatment process. 40 C.F.R. § 260.10.

78.   A by-product is defined under RCRA as a material that is produced in a production process, but which is neither the primary product of the production process, or a co-product intended for sale. Examples are slags and distillation column bottoms. *Id.* § 261.1(c)(3).

79.   Under RCRA, EPA has compiled a list of chemicals that are defined to be hazardous wastes when they are disposed of. *See id.* § 261.33.

80.   *Id.* § 261.2.

81.   *Id.* § 261.1(c)(4).

82.   *See id.* §§ 261.30-261.32 (listed hazardous wastes).

83.   *See id.* § 261.4.

84.   Currently, this exclusion applies only to reclaimed products returned to the original process as feedstock. EPA has, however, proposed a rule to allow any use of the reclaimed material in the original process. *See* 53 Fed. Reg. 519 (Jan. 8, 1988).

85.   40 C.F.R. § 261.6(a)(3).

86.   *Id.* § 261.6(a)(2).

87.   *Id.* Part 266.

88.   *Id.* § 261.6(b).

89.   *Id.* § 261.6(c).

90.   *Id.*

91.   *See* National Governors' Association, *The Role of Waste Minimization* at 14 (1989) (hereinafter cited as "Waste Minimization").

92.   42 U.S.C. § 6924(b).

93.   *Id.* § 6929.

94.   19 Env't Rep. (BNA) 704 (1988).

95.   42 U.S.C. § 9604(c)(9).

96.   Waste Minimization, *supra* note 91, at 76.

97.   *Id.* at 57-85.

98.   1988 Hearings, *supra* note 12, at 30 (statement of Rep. Claudine Schneider).

99.   Waste Minimization, *supra* note 91, at 77.

100.   *See* 1988 Hearings, *supra* note 12, at 280-88 (statement of James Gutensohn, Commissioner, Massachusetts Department of Environmental Management).

101.   National Academy of Sciences, *Reducing Hazardous Waste Generation* (National Academy Press, 1985).

102.   *Cutting Chemical Wastes*, INFORM, 1986.

103.   Serious Reduction, *supra* note 11; Pollution to Prevention, *supra* note 11.

104.   EPA Waste Minimization Report, *supra* note 46.

105.   1988 Hearings, *supra* note 12, at 293 (summary description of H.R. 2800 prepared by the Northeast-Midwest Coalition).

106.   42 U.S.C. § 11023.

107.   Toxics Release Inventory, *supra* note 3.

108.   *Id.* at 14.

109.   *Id.*

110.   42 U.S.C. § 11023(b).

111.   *Id.* § 11023(f).

112.   H.R. 2800 defined *waste reduction* to mean essentially the same as the term *source reduction* as used in this book.

113.   1988 Hearings, *supra* note 12, at 260 (statement of Mort T. Mullins for CMA); *id.* at 250 (statement of V. Wayne Roush for API).

114.   *Id.* at 250-59.

115.   *Id.* at 260-62.

116.   *See*, e.g., *id.* at 139-85 (statement of Gerald V. Poje for the National Wildlife Federation); *id.* at 274-75 (statement of National Association of State PIRGs).

117.   *Id.* at 81-91 (statement of Joel Hirschorn).

118.   *Id.* at 94-96 (statement of Roger Schecter for National Roundtable of State Reduction Programs).

119.   *See, e.g., id.* at 25-31 (statement of Rep. Claudine Schneider); *id.* at 292 (statement of Gov. Thomas H. Kean, New Jersey, for Coalition of Northeastern Governors).

120.   *Id.* at 263-64 (statement of William Y. Brown).

121.   *Id.* at 264 (statement of Jerry B. Martin).

122.   *Id.* at 186 (statement of Jane L. Bloom).

123.   *Id.* at 210.

124.   134 Cong. Rec. H9247 (daily ed. Sept. 30, 1988).

125.   H.R. 1457, 101st Cong., 1st Sess. (1989).

126.   S. 585, 101st Cong., 1st Sess. (1989).

127.   1989 Hearings, *supra* note 49.

128.   1988 Hearings, *supra* note 12, at 32-47.

129.   1989 Hearings, *supra* note 49, at 63 (statement of Robert L. Dostal).

130.   *Id.* at 137-38 (statement of William Y. Brown).

131.   *See id.* at 64-74 (statement of Morton L. Mullins).

132.   *Id.* at 81-84 (statement of William J. Mulligan).

133.   *Id.* at 26, 28.

134.   *See id.* at 25, 31; 54 Fed. Reg. 3845 (Jan. 26, 1989); *see also* p. 61.

135.   54 Fed. Reg. 3845 (Jan. 26, 1989).

136.   1989 Hearings, *supra* note 49, at 25, 31.

137.   *Id.* at 32.

138.   *Id.* at 33.

139.   *Id.* at 28.

140.   *Id.* at 30.

141.   *See* pp. 87-89.

142.   *See* p. 61.

143.   53 Fed. Reg. 27077 (Jul. 18, 1988).

144.   Telephone interview with Jackie Krieger, Program Policy Analyst, EPA Office of Pollution Prevention (Mar. 15, 1989).

145.   42 U.S.C. § 6962.

146.   54 Fed. Reg. 7329 (Feb. 17, 1989).

147.   42 U.S.C. § 6962(e).

148.   40 C.F.R. Part 249.

149.   40 C.F.R. Part 250.

150.   40 C.F.R. Part 252.

151.   53 Fed. Reg. 46558 (Nov. 17, 1988) (to be codified at 40 C.F.R. Part 253).

152.   54 Fed. Reg. 7328 (Feb. 17, 1989) (to be codified at 40 C.F.R. Part 248).

153.   *See* Solid Waste Dilemma, *supra* note 12, at 52.

154.   *See*, e.g., 40 C.F.R. § 252.21-252.24.

155.   *See* 54 Fed. Reg. at 7329.

156.   *See* 54 Fed. Reg. 3845 (Jan. 26, 1989).

157.   54 Fed. Reg. 52209, 52215 (Dec. 20, 1989).

158.   *Id.* at 52213.

159.   *Id.* at 52215, 52216.

160.   *Id.* at 52241.

161.   Fla. Stat. Ann. § 403.708(10) (West Supp. 1989); *see also* 19 Env't Rep. (BNA) 344 (Current Developments, July 8, 1988).

162.   Fla. Stat. Ann. § 403.708(11)(b); *see also* 19 Env't Rep. (BNA) 344.

163.   Portland, Or., Ordinance No. 161573 (1989); *see also* 19 Env't Rep. (BNA) 2074 (Current Developments, Feb. 3, 1989).

164.   *Washington Post*, Feb. 1, 1990, at B-3, col. 1 (legislation to ban or tax disposable diapers pending in California, Connecticut, New York, Ohio, Oregon, Washington, and Vermont).

165.   Telephone interview with David Newton, Recycling Coordinator, Suffolk County, New York (March 15, 1989).

166.   Solid Waste Dilemma, *supra* note 12, at 2.A-7.

167.   Minn. Stat. Ann. § 325E.042 (West Supp. 1989).

168.   1988 Conn. Legis. Serv. P.A. 88-231, § 14 (West).

169.   Minn. Stat. Ann. § 325E.045 (West Supp. 1989).

170.   Fla. Stat. Ann. § 403.7197 (West Supp. 1989); *see also* Solid Waste Dilemma, *supra* note 12, at 2A-11.

171.   Fla. Stat. Ann. § 403.7195; *see also* Solid Waste Dilemma, *supra* note 12, at 2A-11.

172.   C. Peterson, "A Look at Current Waste Reduction," *Waste Age Magazine*, Vol. 20, No. 3, at 11 (March 1989) (hereinafter cited as "Peterson").

173.   R.I. Gen. Laws § 23-18.11-3 (Supp. 1988); *see also* Peterson, *supra* note 172, at 13.

174.   National Solid Wastes Management Association, *Recycling in the States: Update 1989* (Nov. 15, 1989) (hereinafter cited as "NSWMA Update").

175.   N.J. Stat. Ann. 13:1E-99.13 (West Supp. 1988).

176.   Telephone interview with Steve Rinaldi, New Jersey Department of Environmental Protection (March 15, 1989).

177.   New Jersey Department of Environmental Protection, Office of Recycling, "Recycle: It's the Law" (informational pamphlet).

178.   1987 Conn. Acts 544, § 3 (Reg. Sess.); Conn. Dept. of Env. Protection, Mandatory Recycling Regulations § 22a-241b-2 (effective Feb. 28, 1989) (to be codified at Conn. Agencies Regs. § 22a-241b-2).

179.   Conn. Gen. Stat. Ann. § 22a-227 (West Supp. 1988).

180.   Telephone interview with Jackie White, Recycling Specialist, Connecticut Department of Environmental Protection (March 29, 1989).

181.   1988 Pa. Legis. Serv. Act 101, § 1501 (Purdon); *see also* 19 Env't Rep. (BNA) 406 (Current Developments July 22, 1988).

182.   1988 Pa. Legis. Serv., Act 101, § 1501(c) (Purdon); *see also* 19 Env't Rep. (BNA) 406.

183.   Telephone interview with Evelyn D'Elia, Pennsylvania Department of Environmental Resources (March 17, 1989).

184.   1988 N.Y. Laws ch. 70, § 23(2) (to be codified at N.Y. Gen. Mun. § 120-aa(2)).

185.   Oreg. Rev. Stat. § 459.165 et seq. (1987).

186.   Oreg. Rev. Stat. § 459.165(a).

187.   Telephone interview with Lissa Weinhold, Recycling Specialist, Oregon Department of Environmental Quality (March 24, 1989).

188.   Va. Regs. Reg. §§ 2.2, 2.3(3).

189.   Fla. Stat. Ann. § 403.706(4) (West Supp. 1989).

190.   1988 Ill. Legis. Serv., P.A. 85-1198, § 4(a) (West) (to be codified at Ill. Ann. Stat. Ch. 5, ¶ 5954); *see also* 19 Env't. Rep. (BNA) 772 (Current Developments, Sept. 2, 1988).

191.   1988 Ill. Legis. Serv., P.A. 85-1198, § 6(3) (West) (to be codified at Ill. Ann. Stat. Ch. 85, ¶ 5956(3)); *see also* 19 Env't. Rep. (BNA) 772.

192.   Mich. Stat. Ann., P.A. No. 6, § 13.29(30a) (Callaghan Current Legis. 1988); *see also* 19 Env't Rep. (BNA) 1915 (Current Developments, Jan. 27, 1989).

193.   Telephone interview with Steve Kratzer, Michigan Department of Natural Resources, citing the Clean Michigan Fund's Recycling Feasibility Studies of 1986.

194.   *See* NSWMA Update, *supra* note 174.

195.   *See* Del. Code Ann. tit. 7, § 6401 *et seq.* (1983).

196.   Solid Waste Dilemma, *supra* note 12, at 2.B-41.

197.   19 Env't Rep. (BNA) 1436 (Current Developments, Nov. 11, 1988).

198.   Telephone interview with Mr. Charles Baxter, Massachusetts Department of Environmental Quality Engineering (March 20, 1989).

199.   6 N.Y.C.R.R. Part 360-1.9(f) (1988).

200.   19 Env't Rep. (BNA) 1786 (Current Developments, Dec. 30, 1988).

201.   Mich. Stat. Ann., P.A. No. 415 § 13.29(73) (Callaghan Current Legis. 1988); *see also* 19 Env't Rep. (BNA) 1914 (Current Developments, Jan. 27, 1989).

202.   1988 Ill. Legis. Serv., P.A. 85-1197, § 1 (West) (to be codified at Ill. Ann. Stat. ch. 127, ¶ 133b10); *see also* 19 Env't Rep. (BNA) 772 (Current Developments, Sept. 2, 1988).

203.   Fla. Stat. Ann. § 403.7065 (West Supp. 1989); *see also* 19 Env't Rep. (BNA) 344 (Current Developments, July 8, 1988).

204.   Mich. Stat. Ann., P.A. No. 414 § 13.28(82) (Callaghan Current Legis. 1988); *see also* 19 Env't Rep. (BNA) 1914 (Current Developments, Jan. 27, 1989).

205.   1988 Ill. Legis. Serv., P.A. 85-1198, § 10 (West) (effective Jan. 1, 1991) (to be codified at Ill. Ann. Stat. ch. 85, ¶ 15960); *see also* 19 Env't Rep. (BNA) 772 (Current Developments, Sept. 2, 1988).

206.   Solid Waste Dilemma, *supra* note 12, at 2.B-12.

207.   *Id.* at 2.B-11.

208.   S. 2773, 100th Cong., 2d Sess., 134 Cong. Rec. S12172 (daily ed. September 9, 1988).

209.   S. 1113, 101st Cong., 1st Sess., 135 Cong. Rec. S6022 (daily ed. June 1, 1989) (hereinafter cited as "S. 1113").

210.   S. 1112, 101st Cong., 1st Sess., 135 Cong. Rec. S6016 (daily ed. June 1, 1989) (hereinafter cited as "S. 1112").

211.   H.R. 3735, 101st Cong., 1st Sess., introduced November 19, 1989 (hereinafter cited as "H.R. 3735").

212.   H.R. 2845, 101st Cong., 1st Sess., introduced June 29, 1989 (hereinafter cited as "H.R. 2845").

213.   S. 1113, *supra* note 209, § 301(a); S.1112, *supra* note 210, § 104.

214.   The Baucus bill specifies that these percentages are to be achieved by the end of four and ten year terms from the date of enactment of the legislation. The Chafee bill specifies deadlines of 1993 and 1999—four and ten years from the date of introduction.

215.   Section 303 of the Baucus bill requires that in the preceding fiscal year any state seeking a grant have made "satisfactory progress in promoting the use of waste reduction and recycling techniques" that it is pursuing under its solid waste management plan. Section 115(c) of the Chafee bill requires simply that a state have, within five years after enactment of the legislation, an EPA approved solid waste management plan that it is implementing appropriately.

216. H.R. 3735, *supra* note 211, § 801(a).

217. Senator Baucus introduced similar legislation in the 100th Congress that would impose a fee on "virgin material used for packaging" of $7.00 per ton of packaging material or $0.007 per rigid container. *See* S. 2774, 100th Cong., 2d Sess., 134 Cong. Rec. S12182 (daily ed. September 9, 1988).

218. *See* H.R. 3737, 101st Cong., 1st Sess. § 1, introduced November 19, 1989.

219. *Id.* § 1(a).

220. S. 1112, *supra* note 210, § 115(b).

221. *Id.* § 115(d)(1).

222. S. 1113, *supra* note 209, § 310.

223. 15 U.S.C. § 2605.

224. H.R. 3735, *supra* note 211, at § 106(a).

225. S. 1112, *supra* note 210, § 106(b)(5).

226. *Id.* § 108(a).

227. H.R. 2845, *supra* note 212, § 9(a).

228. S. 1112, *supra* note 210, § 106(b)(5).

229. *Id.* § 106(b)(2). The Luken bill would establish an advisory commission that would study, *inter alia*, "what information should be included on product or packaging labels,"such as the "recyclability of the product, the amount of recycled material in the product or packaging, and the presence of toxic constituents in the product or packaging." H.R. 3735, *supra* note 211, § 604(b)(2).

230. S. 1113, *supra* note 209, § 307(e).

231. S. 1112, *supra* note 210, § 106(b)(3); H.R. 3735, *supra* note 211, § 903.

232. S. 1112, *supra* note 209, § 108(b)(3).

233. H.R. 2845, *supra* note 212, § 9.

234. S. 1112, *supra* note 210, § 110.

235. H.R. 3735, *supra* note 211, § 604.

236. S. 1113, *supra* note 209, § 307.

237.   S. 1112, *supra* note 210, § 113(a).

238.   Franklin Associates, Ltd., *Characterization of Products Containing Lead and Cadmium in Municipal Solid Waste in the United States, 1970 to 2000* (January 1989).

239.   S. 1113, *supra* note 209, § 310.

240.   S. 1112, *supra* note 210, § 109.

241.   H.R. 2845, *supra* note 212, § 10(a).

242.   H.R. 3735, *supra* note 211, § 107.

243.   Another such potential tool is the use of tax credits for those who recycle newspapers and other postconsumer discards. Although none of the four bills discussed here contain such a provision, this concept is being discussed extensively in Congress.

244.   In the preamble to the paper guidelines, EPA stated its view that section 6002 of RCRA in its existing form does not provide "explicit authority to EPA to authorize or recommend payment of a price preference or to create a set aside." 53 Fed. Reg. 23546, 23559 (Jun. 22, 1988). Accordingly, in EPA's view, "a price is 'unreasonable' if it is greater than the price of a competing product made of virgin material." *Id.*

245.   S. 1113, *supra* note 209, § 311; S. 1112, *supra* note 210, §§ 103, 111(a)(2); H.R. 2845, *supra* note 212, § 3; H.R. 3735, *supra* note 211, § 701.

246.   S. 1113, *supra* note 209, § 311.

247.   S. 1112, *supra* note 210, § 111.

248.   H.R. 2845, *supra* note 212, § 3; H.R. 3735, *supra* note 211, § 701.

249.   S. 1112, *supra* note 210, § 111(c); S. 1113, *supra* note 209, § 309.

250.   H.R. 2845, *supra* note 212, § 4.

251.   S. 1113, *supra* note 209, § 308; S. 1112, *supra* note 210, § 111(b); H.R. 3735, *supra* note 211, § 602(a).

252.   H.R. 2845, *supra* note 212, § 5; H.R. 3735, *supra* note 211, § 901.

253.   H.R. 2845, *supra* note 212, § 2; H.R. 3735, *supra* note 211, § 702(a).

254.   S. 1112, *supra* note 210, § 116.

255.   *See id.* § 115(d)(3).

256.   H.R. 3735, *supra* note 211, § 702(a).

257. S. 1113, *supra* note 209, § 303.

258. S. 1112, *supra* note 210, § 115.

259. H.R. 3735, *supra* note 211, § 801(a).

260. S. 1112, *supra* note 210, § 106.

261. S. 1113, *supra* note 209, § 302.

262. *Id.* § 304; S. 1112, *supra* note 210, § 107; H.R. 3735, *supra* note 211, § 603(a).

263. S. 1113, *supra* note 209, § 306.

628.4
JAS

This publication is designed to provide accurate and authoritative information regarding its subject matter. It is sold with the understanding that the publisher is not engaged in rendering legal, accounting, or other professional service. If legal advice or other expert assistance is required, the services of a competent professional person should be sought.—*From a Declaration of Principles Jointly Adopted by a Committee of the American Bar Association and a Committee of Publishers.*

1121431 4

Learning Resources
Centre

ISBN: 1-55840-272-1

Library of Congress Catalog Number: 88-083642

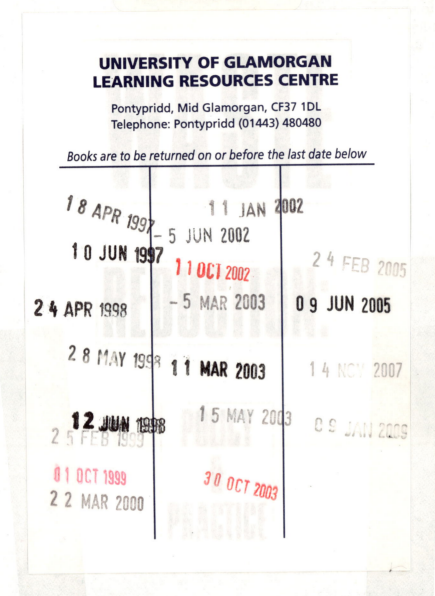
Waste Management, Inc. and Piper & Marbury

Executive Enterprises Publications Co., Inc. New York, New York